Psience Fiction

ALSO OF INTEREST

Evidence for Psi: Thirteen Empirical Research Reports
Edited by Damien Broderick *and*
Ben Goertzel (McFarland, 2015)

Psience Fiction
The Paranormal in Science Fiction Literature

DAMIEN BRODERICK

McFarland & Company, Inc., Publishers
Jefferson, North Carolina

LIBRARY OF CONGRESS CATALOGUING-IN-PUBLICATION DATA

Names: Broderick, Damien, author.
Title: Psience fiction : the paranormal in science fiction literature / Damien Broderick.
Description: Jefferson, North Carolina : McFarland & Company, Inc., Publishers, 2018. | Includes bibliographical references and index.
Identifiers: LCCN 2018016856 | ISBN 9781476672281 (softcover : acid free paper) ∞
Subjects: LCSH: Science fiction—History and criticism. | Paranormal fiction—History and criticism. | Parapsychology and philosophy.
Classification: LCC PN3433.6 .B76 2018 | DDC 809.3/8762—dc23
LC record available at https://lccn.loc.gov/2018016856

BRITISH LIBRARY CATALOGUING DATA ARE AVAILABLE

ISBN (print) 978-1-4766-7228-1
ISBN (ebook) 978-1-4766-3197-4

© 2018 Damien Broderick. All rights reserved

No part of this book may be reproduced or transmitted in any form or by any means, electronic or mechanical, including photocopying or recording, or by any information storage and retrieval system, without permission in writing from the publisher.

Front cover image © 2018 CSA-Printstock/iStock

Printed in the United States of America

McFarland & Company, Inc., Publishers
  Box 611, Jefferson, North Carolina 28640
  www.mcfarlandpub.com

To the memory of John Wood Campbell
and the creators of psience fiction

Acknowledgments

I am grateful to the following rights holders for allowing me to cite several of the major works of what I dub "psience fiction":

For permission to quote very extensively from editorials and letters by John W. Campbell, thanks to John Betancourt of Wildside Press, manager of the Literary Estate of John W. Campbell, Jr.

For permission to quote from *Jack of Eagles*, thanks to the estate, to the Heather Chalcroft Literary Agency and the Virginia Kidd Agency.

For permission to quote from two novels and a short story by Robert Silverberg, thanks to Mr. Silverberg.

For permission to quote from correspondence, thanks to former Star Gate remote viewer Joseph McMoneagle.

For permission to cite his valuable article on psi and sf, thanks to Professor Peter Lowentrout.

Finally, I thank Alec Nevala-Lee, who generously allowed me to read in advance of publication his excellent book on John W. Campbell and that pivotal editor's early stable of sf writers. Titled *Astounding: John W. Campbell, Isaac Asimov, Robert A. Heinlein, L. Ron Hubbard, and the Golden Age of Science Fiction*, it is both enjoyable and a solidly researched work of literary and cultural history.

Table of Contents

Acknowledgments	vi
Preface	1
Introduction	7

1. Donald Macpherson (George Humphrey), *Go Home, Unicorn* (1935) — 21
2. Olaf Stapledon, *Odd John* (1935) — 24
3. E. E. Smith, *The History of Civilization* (1937/1951) — 28
4. A. E. van Vogt, *Slan* (1940/1946/1968) — 32
5. James Blish, *Jack of Eagles* (1949/1952) — 37
6. James H. Schmitz, *The Witches of Karres* (1949/1966) — 42
7. Alfred Bester, *The Demolished Man* (1952) — 45
8. Zenna Henderson, *The People* (1952) — 50
9. J. T. McIntosh, *The ESP Worlds* (1952) — 52
10. Theodore Sturgeon, *More Than Human* (1953) — 55
11. Mark Clifton and Frank Riley, *They'd Rather Be Right* (1953–1956) — 59
12. Mark Clifton, "What Thin Partitions" to "Remembrance and Reflection" (1953–1958) — 63
13. Wilson Tucker, *Wild Talent* (1954) — 67
14. James H. Schmitz, *The Ties of Earth* (1955) — 69
15. John Wyndham, *The Chrysalids/Re-Birth* (1955) — 72
16. Robert A. Heinlein, *Time for the Stars* (1956) — 73

17. Frank M. Robinson, *The Power* (1956) and *Waiting* (1999)   77
18. George O. Smith, *Highways in Hiding* (1956)   80
19. Alfred Bester, *The Stars My Destination* (1956–57)   83
20. Lan Wright, *A Man Called Destiny* (1958)   87
21. Marion Zimmer Bradley, *Darkover* Series (1958–)   90
22. Jack Vance, "Parapsyche," "The Miracle Workers" and "Telek" (1958)   92
23. Short Stories (1940s–1950s)   98
    Katherine MacLean, *"Defense Mechanism" (1949)*   98
    C.M. Kornbluth, *"The Mindworm" (1950)*   99
    Walter M. Miller, Jr., *"Command Performance" (1952)*   100
    Isaac Asimov, *"Belief" (1953)*   101
    Algis Budrys, *"Riya's Foundling" (1953)*   103
    Cordwainer Smith, *"The Game of Rat and Dragon" (1955)*   104
    Brian W. Aldiss, *"Psyclops" (1956)*   105
    J.T. McIntosh, *"Empath" (1956)*   106
    Poul Anderson, *"Journeys End" (1957)*   107
24. Mark Phillips (Randall Garrett and Laurence M. Janifer), *Brain Twister*, *Impossibles* and *Supermind* (1959–1961)   108
25. Arthur Sellings, *Telepath* (1962)   112
26. Keith Woodcott, a.k.a. John Brunner, *Crack of Doom/The Psionic Menace* (1962–1963)   115
27. John Brunner, *Telepathist/The Whole Man* (1964)   118
28. Dan Morgan, *The Sixth Perception* Series (1967–75)   121
29. Richard Cowper, *Breakthrough* (1967)   127
30. Anne McCaffrey, *Talents Universe* (1968–)   130
31. Philip K. Dick, *Ubik* (1969)   133
32. Colin Wilson, *The Philosopher's Stone* (1969)   138
33. Joanna Russ, *And Chaos Died* (1970)   142
34. Lester del Rey, *Pstalemate* (1971)   145
35. Robert Silverberg, *Dying Inside* (1972)   149
36. Katherine MacLean, *Missing Man* (1975)   152
37. Robert Silverberg, *The Stochastic Man* (1975)   155
38. Octavia Butler, *Mind of My Mind* (1976)   158

39. Joan D. Vinge, *Psion* (1982)	162
40. Lucius Shepard, *Life During Wartime* (1987)	164
41. Carrie Vaughn, *After the Golden Age* (2011) and *Dreams of the Golden Age* (2013)	167
42. Connie Willis, *Crosstalk* (2016)	169
43. Two Novels by Psychics (1978, 1998)	173
44. Short Stories (1960s–1990s) Poul Anderson, *"Night Piece" (1961)* 178 Robert Silverberg, *"Something Wild Is Loose" (1971)* 180 C.J. Cherryh, *"Cassandra" (1978)* 181 Brian M. Stableford, *"The Oedipus Effect" (1991)* 182	178
Conclusion	184
Appendix 1: A Brief Guide to Paranormal Research	195
Appendix 2: Psi and Afterlife in Psience Fiction	207
Chapter Notes	218
References	223
Index	229

Preface

Three major icons or narrative devices dominated what is now known as the Golden Age of Science Fiction (sf[1]), from the late 1930s to the end of the 1940s and arguably through the 1950s. These were

(1) the "conquest" of space,
(2) command of atomic nuclear forces for war and peace, and
(3) ESP and other psychical phenomena, sometimes dubbed "psionics."

Prior to Hiroshima and then the Apollo 11 lunar landing, all three of these major icons were widely regarded by non-sf readers as intrinsically ridiculous although perhaps entertaining in a childish way, akin to heroic stories of knights and enchanted princesses and dragons. It was the sort of thing the kids might enjoy after they completed their homework and sports commitments. Any adults who continued reading the gaudy magazines and paperbacks containing such stories tended to be scorned.

Then things changed—literally with a bang. With the passing of very few decades, and the visibly accelerated pace of science and technology, the first two crazy icons became an authentic and even frightening reality.

The psychic element, though, remained as implausible as magic or Atlantis or astrology (although millions continued to consult supposed uncanny influences from "the stars").

$$\Psi$$

Why do so many people believe in psychic (or psi) phenomena, or at least in one or other of the smorgasbord of available paranormal mysteries? Such purported abilities, apparently at odds with science, range from telepathy between twins, and dogs that know when you're coming home, to frightening visions of future disasters, and even to bilocation and the appearance of jewelry and other objects out of thin air. All of these claimed events and powers do certainly *seem* inconsistent with the known and well-established laws of physics and biology. That's why popular online sources of information such as Wikipedia evince little caution in declaring parapsychology as being "identified as a pseudoscience by the overwhelming majority of mainstream scientists" ("Parapsychology," as of July 7, 2017). It used to open even more bluntly: "Parapsychology is a pseudoscience." The entry continues:

> the subject rarely appears in mainstream science journals. Most papers about parapsychology are published in a small number of niche journals. Parapsychology has been criticised for continuing investigation despite being unable to provide convincing evidence for the existence of any psychic phenomena after more than a century of research.
>
> It has been noted [by the physicist Sean Carroll, in *Wired*, 2016] that most academics do not take the claims of parapsychology seriously.

Actually, there is abundant evidence for psi's reality. Some of it, for readers unfamiliar with the topic, is provided in Appendix 1. For many years there have been university chairs and lectureships dealing with ESP and other psi phenomena in the UK, U.S., Germany, Holland, Sweden and other nations, so this claim that parapsychology is a pseudoscience is rather peculiar, a *parti pris*. To my knowledge there are no professorial chairs in Easter Bunny Research or Santa Claus Studies or Creation Science or Flat Earth Geology.

This book focuses on the way in which, for a time—especially in the 1950s—ESP was the hottest trope in the broad fields of science fiction.

What most readers of these thrilling tales in the mid 20th century did not realize was that such uncanny phenomena might, after all, be genuine, unlike Superman's ability to fly or Wonder Woman's to bounce bullets off her bracelets. Yet the capacity of science fiction to render strange mental abilities believable very likely played a significant role in persuading some Enlightenment-trained skeptics to wonder. Were these vividly imagined phenomena possibly real after all, once the exaggeration due to wishful thinking was peeled away from the fantastic tales? If so, might they be worth investigating with government funding?

The answer, it turned out, just like the reality of orbital spacecraft and Moon landings, and nuclear explosives or power reactors, was Yes.

Ψ

ESP stands for *extrasensory perception*, or as it was hyphenated in the mid–1930s by its originator—the botanist and psychologist Dr. Joseph Banks Rhine—*extra-sensory perception*. That term covered at least three purported paranormal skills.

One was *clairvoyance* (French for "clear seeing"): the non-inferential experience of perceiving or knowing events that occur beyond the range of normal sight or hearing, and without the use of instruments such as telephone, radio or even smoke signals.

Another was *telepathy*, the uncanny experience of "mind reading" or "thought transference" between two or more people also cut off from normal or machine-mediated possibilities of communication. A formal double-blind variety of ESP, developed by physicists Russell Targ and Hal Puthoff for the long-classified psychic program of the U.S. government known most recently as Star Gate, is dubbed "remote viewing."

Both those kinds of *paranormal phenomena* resembled vision (or hearing, for that matter, in cases known as *clairaudience*), but without the intervention of organic eyes or ears. That's where the "extra" part of the term "extrasensory" comes from. It doesn't mean, as many people suppose, that such purported paranormal abilities involve an *extra* or *additional* sense (as implied by the fabled but erroneous term "sixth sense").

Actually, Rhine meant (and wrote explicitly) that the "extra" part of the term just meant that these phenomena were *not* sensory but were *beyond* the senses. We're familiar with this meaning from words like "extraordinary"—which means "far from ordinary," not "really really ordinary"—or "extramarital sex," which does not mean "lots of additional whoopee with your spouse," but rather "having sex with someone you are not married to."

So "extrasensory perception" was an unfortunate choice, because it does seem to imply a dubious theory of how ESP works: that there's a specialized but unidentified organ in the brain or nervous system, never detected by autopsy or subtle medical scanners.

That discarded term "sixth sense" originally had another meaning, the third kind of ESP: prophecy of the future, eerie premonitions especially of doom or disaster, based not on inference from knowable evidence but purely on inexplicable intuition or hunches. Studied in the lab, often in dreary endless efforts to predict a future sequence of cards or sequence of lights on a random number generator, this was termed *precognition*.

In addition to these quasi-perceptual anomalies, folklore tells of a quite different kind of paranormal ability, or happening, that is more akin to muscular lifting, throwing, writing, or even floating free of the force of gravity. This used to be called "mind over matter," or more formally, *telekinesis*, but Rhine and his colleagues preferred their own variant: *psychokinesis*.

Partly, I suspect, this change in terminology was meant to sanitize the previous associations with spiritualism and mediumship, where the persistent dead were supposedly responsible for lifting heavy tables and other objects in often darkened rooms while mysterious trumpets or accordions played. Rhine and others regarded their own work as scientific, with an emphasis on ever-more careful controls, repeated trials, and statistical examination of the attempted effects. Since this sort of approach was then the domain of the newish science of psychology, *psycho-* replaced *tele-*. By an unfortunate coincidence, "psycho" happened to enter slang as a substitute for "lunatic" or "deranged," which pleased the skeptics who found the whole idea of mind-reading and especially mind-over-matter to be ludicrous, a deplorable medieval relic.

More recently, terminology even more disinfected has been urged: *anomalous cognition* and *anomalous perturbation*, for example. One rather less clunky if equally non-obvious usage took hold in 1942, encapsulating the entire range of such alleged deviations from mechanical cause-and-effect. This was the term *psi*, coined by Drs. Robert Thouless and Bertold Wiesner. "Psi" is now almost universally accepted as the general term for both perceptual and motor paranormal anomalies, implying a unitary basis for these apparently quite distinct phenomena.

As we will see, it took on further coloration with the coinage *psionics*— particularly in science fiction, where "psionics" conveyed a sort of not-yet-developed psi equivalent of electronics—psi-powered machines, sometimes of a purely symbolic kind with no actual energy supply or working parts (such as the rods used by dowsers). That's getting awfully close to magic, and some parapsychologists do argue that magic was, indeed, essentially an early attempt to train and constrain psi activities, which then got cluttered up and contaminated with a whole mess of superstitious nonsense, the sort of thing used to justify human sacrifice and the murder of witches and seers accused of willing possession by demons.

Although regarded by many as entirely bogus, the domain of 19th century mediums and swamis and spiritists, such alleged phenomena rose up anew like a narrative tsunami in the science fiction magazines of the 1940s and 1950s, then diminished from the fiction of strange science without ever going away.

The result, in its heyday, was a quite new and startling variety of science fantasy: an imaginative literary exploration (and thematic exploitation) of the paranormal. For example, H.L. Gold, the editor of *Galaxy*, made paranormal-fiction history with two dazzling, baroque, tectonic serials by Alfred Bester: *The Demolished Man* and *The Stars My Destination* (known also as *Tiger! Tiger!*, from the poet William Blake's brightly burning verse). The latter was reviewed by *New World* magazine's Leslie Flood as "packing into the story practically every device known to 'psience-fiction,' plus a few original twists of his own."[2] (I find Flood's coinage so wryly apt that I have borrowed it for the title of this volume.) He concluded that the novel "must surely take its place among the top ten science-fiction novels of all time," an estimate echoed in Samuel R. Delany's assessment that it is "considered by many readers and writers, both in and outside the field, to be the greatest single SF novel. In this book, man, intensely human yet more than human, becomes, through greater acceptance of his humanity, something even more" (Delany, 1977, p. 14). This is the theme of transcendence that, as we shall see, can be regarded as the very spine of psience fiction.

An important analysis by Professor Peter M. Lowentrout (1989) noted some three decades ago the ebbing popularity of psi as a major component of science fiction by that time (and hence, directly, a waning of commercial support for psience fiction *per se*). He drew attention, however, to shifts in "the way the genre's authors use psi ... and the steady growth of the genre as resulting in part from its functioning as a displacement of religious concerns." Drawing on his understanding of world and text from a religious studies perspective, he emphasized the theme of transcendence and its capacity to represent and engage certain ways of thinking and feeling that had been steadily more and more disregarded by "literary" fiction. We shall return to his perspective after examining a significant sample of psience fiction texts.

Ψ

Even after the trope had lost its first feverish appeal to increasingly jaded fans, while perhaps infiltrating much of the metaphysical unconscious of the mode, certain paranormal abilities were being studied and developed for real under top secret security cover for twenty years in the United States and the Soviet Union. That is, real scientists were busily studying real paranormal abilities, eventually with the long-classified budget support of CIA and DIA and the U.S. Army.

In a military program run out of a pair of rather dilapidated houses in the grounds of Fort Meade, Maryland, soldiers were taught to cast their mind's eye into the nowhere to find and describe targets unobtainable by ordinary means, including spies and downed aircraft and canny estimates of the situation.[3] This was the true world of ESP. In some ways it had been foreseen by some sf writers, while in many other respects there was barely any recognizable affinity.

That's what we'll be looking at in this book: the give and take between the science fiction paranormal and the real-world kind, tested and tallied by psi researchers. Of course, hardly anybody gave these parapsychologists any credence, except for those in the military (in the U.S. and the Soviet Union, certainly, as we know now that the programs have ostensibly been shut down and their documents declassified, and by Japanese companies and Chinese research teams, and probably in other nations as well) and all the hundreds of millions or maybe billions who believed that the paranormal was normal and everyday, however spooky and inexplicable it seemed.

Necessary Warning, Without Apology

Because this book is a study of the interaction between scientific studies of paranormal phenomena and their representation in science fiction, it discloses many details of the novels it investigates—and could not do otherwise.

So, be warned: *SPOILERS ARE HERE.*
LOTS OF SPOILERS.
THIS BOOK IS MADE OF SPOILERS!

Readers who insist on finding out "what happens next" in the usual manner should consider reading each novel before proceeding to the chapter discussing it.

Introduction

This book, then, is about perhaps the strangest aspect of science fiction: its long devotion, peaking in the 1950s but prevailing to some extent even now, to *psi*, or *psionics*. "Psi" is not an abbreviation (like ESP for *extrasensory perception*), each letter standing for a complete word. It's simply the 23rd letter of the Greek alphabet, Ψ, pronounced the same as "sigh," and in English should not be capitalized, nor the "p" voiced—although, because of some odd military tradition, the remote viewers of the Star Gate program did often mistakenly spell it PSI, all caps. This led some people to pronounce it Pee Ess Eye, as in the engineering shorthand psi, for Pounds per Square Inch. Wrong. Sigh!

The splendid and comprehensive *Encyclopedia of Science Fiction* (3rd edition, online) declares boldly: "Flight into space is *the* classic theme in sf," adding: "It is natural that sf should be symbolized by the theme of space flight, in that it is primarily concerned with transcending imaginative boundaries, with breaking free of the gravitational force which holds consciousness to a traditional core of belief and expectancy."

Arguably, the most effective prod to young scientists and engineers, filling them with palpable dreams of space, were the science fiction tales of Robert A. Heinlein and Arthur C. Clarke. The titles of their carefully worked-out space adventures tell the story: *The Man Who Sold the Moon*, *The Green Hills of Earth*, and young adult masterpieces such as *Space Cadet*, *Red Planet*, *Starman Jones*, *Between Planets*, *Time for the Stars*, and *Have Spacesuit—Will Travel*, and the script of the 1950 movie *Destination Moon*, all by the American former naval junior lieutenant Heinlein, and *Prelude to Space*, *The Sands of Mars*, *Islands in the Sky*, *A Fall of Moondust*, and, a year before the first real moon landing, the mythic Kubrick movie *2001: A Space Odyssey*, by British wartime radar experimenter Clarke.

And behind these two brilliant polymaths—you wouldn't dare call them "geeks," even if the word had then meant what it's taken to mean now, mockery mixed with grudging respect and envy—was the burly, bristle-headed editor of *Astounding Science Fiction* magazine. John Wood Campbell, Jr., was a powerhouse of futuristic and sideways notions. He developed the best ways to capture and convey these ideas in fiction for engineers young and old, and for many other budding space cadets as well.

Editor Campbell created the golden age of science fiction by creating a new kind of rough-edged literature crammed with extraordinary ideas and stories told as if for the people of the future to read. People of the present day had to fill in all the gaps from the hints and meaningful absences in the richly imagined texts.

That was largely Heinlein's doing, under the guidance of editor Campbell, quickly adopted by youthful genius Isaac Asimov, the Canadian technomystic A. E. van Vogt (later a sucker for L. Ron Hubbard's Dianetics, which also got its start in the pages of *Astounding*), and dozens of others. It was not the whole of science fiction, but in certain respects it was the core, the axle, the best-paying and best-selling and most attractive of these low-rent magazines.

Until it wasn't, any longer, after more literate and urbane competitors arrived at the start of the 1950s: *Galaxy*, home of sardonic satire, and the *Magazine of Fantasy and Science Fiction* (aka *F&SF*), and plenty of others that were mostly gone by the end of the decade when magazine fiction was eaten by television and strange legal maneuvers. But for a time, *Astounding* was the place to go for stories of space travel and starships.

Ψ

And there were atoms—for power, and for nuclear weapons so dreadful that (even though Herbert George Wells had predicted a kind of "atomic bomb" as far back as 1914) few non-experts took the idea as seriously as John Wood Campbell, Jr., and his crew of techno-savvy storytellers.

Damon Knight, a brilliant sf writer, editor and critic, reported sardonically: "Campbell ... told me he wasn't sure how much longer he would edit *Astounding*. He might quit and go into science. 'I'm a nuclear physicist, you know,' he said, looking me right in the eye" (Knight, 1976, p. 114). As a promoter of L. Ron Hubbard's crackpot mind therapy, Campbell wrote a short chapter for the 1950 volume *Dianetics: The Modern Science of Mental Health* discussing scientific method and signed himself "Nuclear Physicist." He had graduated with a bachelor of science degree in physics from Duke University in Durham, North Carolina, in 1932, when the physics of the nucleus was in its rudimentary childhood. (The first sustained chain reaction in an "atomic pile" would not be achieved by Enrico Fermi in Chicago until December 1942, leading very swiftly to the atomic fission bomb program.)

Nevertheless, the *idea* of immense energies available locked inside the atom had a long lineage, most often explored and exploited in magazine adventure fiction. (This present book will largely abstain from considering the many usually derivative and often dumbed-down movies and television shows enacting the topics discussed here, but an excellent critical analysis and international filmography is *Nuclear Movies* [Broderick, Mick, 1991].

Albert I. Berger (1979) notes astutely:

> As a metaphor, atomic energy filled SF magazines long before the Manhattan Project demonstrated the actual powers released by the split nucleus. Atomic power plants were propelling men from star to star as well as revolutionizing life on Earth. It was the central component of the belief that technological innovation was the principal revolutionary force in the world. Once nuclear energy gave promise of actually fulfilling dreams of unlimited power, boundless social changes could be envisioned. Eventually, this metaphor, enriched by an awareness of the new, real research into the nucleus which characterized physics in the 1930s, was combined with the demand for better and more realistic stories which marked the genre as a whole and *Astounding Science Fiction* (ASF) in particular.

Dr. Berger's essay cites several of the key magazine stories pursuing this path to technological transformation and its dread risk of doom. His first example, published in *Astounding* five years prior to Hiroshima and Nagasaki, is A. E. van Vogt's *Slan*, which brings together the three major thematic elements highlighted in this Introduction, combining:

> nuclear energy, telepathy, space flight, and the persecution of racial [more exactly, psychic] minorities as elements in a complex narrative. As he flees from an angry society patterned after Nazi Germany, young Jommy Cross, a telepathic mutant, is searching both for allies in the political struggle against the persecution of the slans and for clues to his own parentage. The key to both quests is an invention of his dead father's, an atomic power device found hidden in a secret laboratory. He develops the device into a propulsion system for a spaceship which he intends to use as a weapon to overthrow the government.

Many of the stories of the atom were more disciplined than that, and those published after the first two city-wrecking bombs in Japan were bitterly melancholic. Expectations of a Third World War, a final war to end all wars in the ashes of universal defeat, were a threnody in many sf stories in *Astounding* and elsewhere. Theodore Sturgeon's poignant "Thunder and Roses" (1947) portrayed an America reeling under nuclear attack, with its military abruptly uncertain of the morality and utility of a retaliatory volley that might end civilization if not all life. Two years later, Sturgeon explained his motive:

> There is good reason to believe that, outside of the top men in the Manhattan District and in the Armed Forces, the only people in the world who fully understood what had happened on August 6, 1945, were the aficionados of science fiction—the fans, the editors, and the authors. Hiroshima had a tremendous effect on me.... Years before the Project, and before the war, we had used up the gadgets and gimmicks of atomic power and were writing stories about the philosophical and sociological implications of this terrible new fact of life [Sturgeon, 1997, pp. 347–48].

A striking image by artist Hubert Rogers illustrated the serial "Fury" (1947), by Henry Kuttner and Catherine L. Moore writing as Lawrence O'Donnell. It showed a somber parade of worthies in an underwater Keep, on Venus, to which the remnants of humankind have fled, passing an immense globe of our planet atop a marble memorial carved with great letters:

> MAN'S GREATEST ACHIEVEMENT
> A WORLD DESTROYED
> BY ATOMIC FIRE

Kuttner and Moore's final novel, *Mutant*, this time under the name Lewis Padgett and first seen as a series in *Astounding* in 1945–53, took up the idea of genetic changes wrought inadvertently by atomic radiation. Their new species were the Baldies, a telepathic offshoot from *Homo sapiens* treated badly, as usual in these tales, by their envious and frightened predecessors.

This van Vogtian trope spread through the genre. Britain's John Wyndham carried it to a fine literary expression in perhaps his best novel, 1955's *The Chrysalids* (the U.S. variant title is *Re-Birth*). After global nuclear war, the remnants of humanity defend their genetic purity, living by the slogan KEEP PURE THE STOCK OF THE LORD; WATCH THOU FOR THE MUTANT. They righteously destroy animal and plant variants (the Offenses) and banish to the wasteland any humans blighted by radiation (the Blasphemies). The new kind is evolving into telepaths, each a chrysalis ready to release its fragile paranormal wings. Fans of sf often embraced these psychic wounded, seeing in them something heroic and undervalued. "Fans are Slans" became an only slightly self-mocking banner. It could equally have been "Fans are Chrysalids."

Ψ

So here is the strangest element of all: the hunger for paranormal transcendence, sometimes fused with what is now called the transhuman or even the posthuman. Mutants with mindboggling psychic powers. Mutants with telepathic tendrils. Levitators, and men and women who could move things by the force of their personality, and people with the rare gift of shifting from our tiresome reality into alternative universes (this before the Many Worlds Theory was on the lips of every cosmologist). It wasn't fantasy, because it had to operate under some kind of lawful constraints. It was the paranormal treated as a kind of science.

It was ESP and future-telling and more, and then more again: it surged through the sf world like a contagion, driven in part by Campbell's urging and endless teasing editorials, and then by those other editors working adjacent streets of the scientific imagination. *New Worlds*' English editor John Carnell adopted the new craze with a certain diffidence but found new sources for his magazine, some by authors from his own nation, some pilfered from the U.S. The varied editors of *F&SF* offered over the decades from 1952 to 1980 schoolmarmish Zenna Henderson's touching if sometimes mawkish reports on the People, evolved humanoids from space crash-landed on Earth and keeping their virtuous heads down—

Meanwhile, as noted earlier, real scientists began busily studying real paranormal abilities, eventually with the secret budget support of CIA and Defense Intelligence Agency and the U.S. military, in a program run in the grounds of Fort Meade, Maryland, where soldiers were taught to cast their mind's eye into the nowhere to find and describe targets unobtainable by ordinary means.

Ψ

Although I have stressed the crucial role of John W. Campbell in spurring his contributors' interest in psi phenomena, and will return to this topic frequently in what follows, it is important to understand that earlier writers had already been influenced by the possibilities of unusual mental abilities. One of the greatest prognosticators and explorers in early sf, before it was widely known as such (let alone as "sci-fi"), was the British philosopher and Marxist Olaf Stapledon. His magisterial novels, most notably *Last and First Men* (1930), *Odd John* (1935) and *Starmaker* (1937), invoke telepathy as an evolutionary advance to be expected both immediately (as with the mutant superman John Wainwright) and, more elaborately, in the deep future. Consider the mind sciences of artificial human beings and Martians in *Last and First Men*. Among the Martians, telepathy created a group mind, "a single psychical process embodied in the electro-magnetic radiation of the whole race," but this gestalt tended to obliterate the distinctions between people— a feature avoided by the Fifth Men. Their telepathic communication was constrained by individual choice. Yet while retaining independence, they could join directly with others, sharing experiences without the need for clumsy verbal symbols. Disagreements might still arise, but they were readily resolved. "Sometimes the process would be easy and rapid; sometimes it could not be achieved without a patient and detailed 'laying of mind to mind.'"

Telepathic linkages and protracted lifespans meant close friendships were possible with very many others. Yet this did not act to restrict the sense of self, but to enhance it, so each person shared in the species' culture while losing nothing. Stapledon's utopian impulse prompted a vision of this state of psientific aggregation as a marvel of humanistic enterprise. For many millions of years, the Fifth Men culture was fixed "almost entirely on art, science and philosophy, without ever repeating itself or falling into ennui" (Stapledon, 1966, pp. 228–30).

Needless to say, this kind of bald "expository lump" was already superannuated when these novels appeared in the 1930s, increasingly replaced by energetic if rarely nuanced dramatization. Indeed, it is often said that Campbell's ascension to the editorship of *Astounding* is just what forced a paradigm change in the way sf was conceived and written. Of course other editors were not without skill, and by the early 1950s would create in their own magazines a kind of aesthetic advance that simply didn't interest Campbell even when some of his favored writers (Sturgeon, Kuttner and Moore, a few others) proved just as capable. It is not that he went out of his way to purchase crudely composed fiction, or to reject poetic or character-sensitive work if it came his way. Rather, his goal was ideational and increasingly quirky, with a marked taste for engineering-based ingenuity powering a ripping yarn.

Behind, or alongside, this pragmatics of publishing a magazine devoted to lives and worlds of a wholly imaginary cast, Campbell grew increasingly devoted to the study of psi phenomena. At the start of the 1950s he was intensely involved with L. Ron Hubbard's Dianetics, but within a few years his ardor cooled, and his concerns began to center on psi. He was not much interested in the tedious card-guessing experiments of Rhine and his colleagues, nor their attempts to control the fall of tossed dice.

Campbell was after something closer to the practical: what he came to call *psionics*. He hoped to chase down the working principles of those anomalies pursued by Charles Fort (1874–1932), whose books dealt with reports of fish falling from the sky, strange figures menacing the quotidian or ignoring it, lights in the heavens that would later be dubbed UFOs, poltergeists, apports, teleportation and telepathy. These books had resonant titles, appealing to the eccentric: *The Book of the Damned, Lo!, Wild Talents*, and *New Lands*. The key notion of Wild Talents, preferably after they were harnessed by scientific method, is what galvanized Campbell's enthusiasm.

With his second wife, Peg, he spent a remarkable amount of time and effort in this quest. In a long letter to British sf author and Fortean Eric Frank Russell, dated October 1, 1952, he deplored weaknesses in the approaches of both Hubbard and Fort. "[Fort's] data was valid. It contained important understandings, and important clues. In that, he was right. But why didn't *he* do some of the hard work of integrating it and finding the pattern" (*The John W. Campbell Letters*, vol. 1, p. 70).

This was not just the irritation of a born editor reading work that trailed off without a denouement. His intention, which was expressed repeatedly in numerous editorials about psi and stories he drew from his stable of writers, was to get this weirdness under control. He explained to Russell:

> Peg and I have done it. We have the basic understanding of what the psi functions are, and how they work. It took us over two years of damned hard work. The reason why I'm now starting it in the magazine [*Astounding*] is that I do have some integrated understanding of what we're dealing with. I'm not yet ready to say a damned thing about it, either, because I recognize that Fort *was* wrong, and what the right answer is. Until I can demonstrate the phenomena

myself, and communicate the exact nature of the mechanisms involved, with demonstrations of each step, I'm not ready to talk. When I've done that, though, by God the physical scientists *will* gladly pitch in and help. I know the general concept of teleportation, levitation, and a few other spontaneous psi phenomena—also telekinesis, etc. In addition, I know the general basic laws which can permit precognition, and an absolute barrier of pure force that will block passage of *any* force now known to physical science [pp. 70–71].

Granted, this might seem implausible, but Campbell set Eric Frank Russell's mind at rest:

"I am not kidding.
"I am not cracked either" [p. 71]

Well, but if this were so, why didn't Campbell reveal at least some of this advanced knowledge to the world? (As far as I know, it still remains undisclosed.) Because:

These forces are real, and I have a theory of their structure. I haven't developed methods of setting up an experiment however, and until I can demonstrate it at an experimental level, it simply doesn't count....

So, the first step toward getting interest in psionics started is to establish *that there is a reward to be earned*....

Reward for considering that psionic forces are real, and actually constitute a level of force below the sub-nucleonic; amusement, plus a hint of satisfying, yet intriguing, possibility [pp. 71–72].

The kinds of rewards Campbell was suggesting were threefold: intellectual fun of the bull session kind; eventual glory and profit from the application of this new psionic framework to technology and science; and most immediately, the reward of having a story incorporating these ideas accepted by *Astounding Science Fiction*.

Ψ

Intriguingly. Campbell actually specified his embrace of what we are calling psience fiction in a long 1953 letter to Dr. J.B. Rhine. First, he reminded Rhine that he had studied at Duke and indeed had contributed his guesses to a set of ESP tests run in Rhine's department:

I attended Duke, quite some years ago; somewhere in your records must be some of the runs on the ESP cards that I made. Later, for some years I lived across the street from the brother of your experiment designer, Dr. Charles Stewart. I had a good many discussions with Charlie about your work [p. 222].

Most of Campbell's letter is a wandering discussion of logic versus empiricism, arguing that Rhine's emphasis on psychology as the prime discipline for studying psi was misplaced. Physics, this former physicist asserted, was the relevant domain. "Physicists in the 25 to 35 year age bracket are looking for new projects to study. They are, probably, most apt to be willing and competent to search out the new basic laws of the Universe which underlie the psi functions.

"Getting them to do so, however, is something of a trick" (p. 225).

Campbell explained candidly to Rhine the approach he'd earlier proposed to Eric Frank Russell:

> The psycho-socio power of fiction as a medium of communication has been somewhat overlooked and underrated, I believe. Jesus used fiction as one of his most powerful teaching tools.
>
> I am trying to use fiction to induce competent thinkers to attack just such problems as the psi effects; my magazine is widely read by creative, speculative, physical scientists. The students at major universities read it—and so do their instructors.
>
> Currently, I am seeking, through the fiction, to nudge interest in psionic powers as an engineering value [p. 225–26].

Why engineers in particular? Because they were interested in results, not the approval of those narrow-minded theoreticians.

> The theoretician feels satisfied when he has proven to his satisfaction that "What you want to be can't be done. I have proven it is impossible."
>
> The horny handed engineer can be a great trial to such a theoretician. He's apt to go out and do that impossible, forcing the unhappy theoretician to revise all his theories.
>
> In our fiction, therefore, our major attack on the Society's block against the psionic functions is at the level of engineering applications of the psi functions—and acknowledging that they work only statistically. The engineer is quite happy with statistical success, because he can simply use a factor of safety [pp. 227–28].

And here was his psience fiction method of reaching and fertilizing the imagination of those practical engineers:

> The Christian doctrine of "By their fruits ye shall know them" is solidly valid. "Make it work!" is the equivalent statement.
>
> In fiction, I can make it work. Since human entertainment and relaxation is a very important aspect of living—why, I can make the psionic forces work very nicely, right now, at an engineering level.
>
> But there's a sly trick here. If the reader is to enjoy the entertainment of the story, *he must temporarily accept the validity of psionic powers.* Never again can he be *wholly* opposed to the idea, for he has already accepted it in a certain degree. Accepting the idea is already associated with pleasure-satisfaction; that association makes it psychologically difficult for him to reject the idea flatly [pp. 228–29].

Campbell closed with a call for solidarity between the science fiction writers and the parapsychologists: "Give me time, Sir! I'm in your business too!" (p. 229).

What was Rhine's response to this offer of a propaganda wing of the psi explorers and advocates? There is no known record of any reply.

Ψ

So John W. Campbell set out to use his magazine *Astounding Science Fiction* as a vehicle for both the investigation and promotion of psi. In the September 1955 issue, he printed a long letter by one T. O. Jothun (a pseudonym) describing experiments on the effects of microwaves on the human body and mind. Allegedly Jothun had discovered remarkable psychic developments, and he urged Campbell to promote research in this area.

In the February 1956 issue, Campbell reported, in an editorial baldly headed "The Science of Psionics," that this letter "drew more reader response than any other item in the magazine. It's clear that there is a very strong and dynamic interest in the type of material Jothun offered" (p. 6). No editor was likely to ignore such a prompt. In other quarters, lesser magazines were selling extremely well by pushing the reality and dangers of UFOs.

Campbell's lure would be psionics, a merging of paranormal phenomena with mechanisms capable of detecting and perhaps amplifying such effects. But he was at pains to point out that any such investigation, at that point, was necessarily strictly *unscientific*. (He didn't mean *anti*scientific, but rather *pre*scientific.) It lacked theory, and however honest its explorers might be, they could not obtain the kinds of repeatable results available to scientists working in established disciplines. This was a frank and rather disarming admission.

He went on: "But I must state clearly beforehand that the statements made in such articles will be claims of having accomplished things that any intelligent modern man knows are clear, pure nonsense—impossibilities. Precisely; that is the necessary condition for proof of discovery.... It is not *demonstration* that is lacking, but *explanation*" (p. 158). So his approach would differ from that of parapsychologists such as Dr. Rhine. As the editor of the leading magazine filled with speculative fiction, written and read for the enjoyment of testing the limits of the known, he was under no obligation to provide a new theory of the universe capable of including the bizarre phenomena of psi.

In a subsequent editorial in June 1956, "The Problem of Psionics," he was even more explicit: "The only sane thing we can do is say, mentally, "O.K.—so we're fumbling amateurs, and we don't know what we're talking about. But if it works, if it is useful to all, in any way, it's a worthwhile gimmick. And if it never does a darned thing of any practical value—fine. I've had fun trying" (p. 5).

He did not really mean that, of course. In January 1959 (six years after his letter to Rhine), in another quite serious editorial, "We *Must* Study Psi," he noted the already long history of psi in science fiction:

16 Introduction

> During the last four years, I've been investigating psi: I started the investigation largely because it has been a background element in science fiction, almost from the start. Telepathy has been stock business. E.E. Smith's Lensmen series was based primarily on psi—for the Lens itself is, essentially, a psi machine.
> With the development of science into engineering proceeding at the pace it has, by 1950 the major developments that science fiction had been forecasting were definitely under engineering—not theoretical—study. It was time for us to move on, if we were to fulfill a function as a frontier literature.
> To some extent, science fiction moved on into the social sciences—sociology, anthropology and psychology.... I was forced back toward psi, even when science fiction started toward the social sciences [*Astounding*, January 1959, pp. 4–5].

In an earlier and appealingly candid declaration, he had written: "Since I published the editorial in the February 1956 issue, suggesting running material on psi machines, I have been receiving quantities of information, from hundreds of sources" (p. 5). Yes, but aside from this hint to keep publishing fiction and nonfiction on the topic in order to sell more copies, can it really be true that we *must* study psi? Bear in mind that this was still the heyday of Behaviorism, which had apparently trounced psychologists of the unconscious such as Freud and Jung, certain that minds were simple if enormously elaborate machines. Noam Chomsky's devastating review of B.F. Skinner's magnum opus, demolishing that entire research program in a single blow (or so it seemed), would not appear for another year. So within that *Zeitgeist* it was somewhat scandalous, like a confession of metaphysical conversion, for Campbell to write:

> Psi phenomena exist at the same level that emotion, desire, and want do, as far as I can make out. If that's the case, then in studying the psi phenomena, you're studying the level which men, today, hold to be the ultimate level of privacy—Subjective Reality. An understanding of the laws of this level would make it possible to manipulate desire, change attitudes, control emotions....
> I suggest that Subjective Reality bears the same relationship to Objective reality that field-forces do to matter. Field forces are not material; they have wildly different laws—but they do obey laws.
> I suggest that Subjective Reality is a true, inherent level of reality in the Universe [pp. 159–60].

If this is the case, perhaps it makes sense that "*we must study* psi, *because it is the only objectively observable set of phenomena stemming from subjective forces*" (p. 161).

In later issues, Campbell spent time promoting a mysterious gadget he called the Hieronymus machine, patented by one T.G. Hieronymus, a box containing a prism and amplifier tubes and resistors, and yielding curious subjective tinglings at certain dial settings, mostly idiosyncratic for each user. To the inventor's dismay, Campbell let his ad hoc theorizing lead him to a preposterous extension: a symbolic version of the machine with no internal parts, just a circuit diagram. Apparently it worked just as well as the original.

Several years after that, Campbell's enthusiasm shifted to a non-psionic device he called the Dean Machine, another patented gadget that purported in this case to change rotary motion into unidirectional thrust (*Astounding/Analog*, June 1960, pp. 83–106). At this stage, the scientifically trained came baying for his blood. Conservation of momentum, sir! You just can't do that! Was that a grim set to Campbell's mouth, or a grin? Some years later, a young Barry Malzberg came out of a confrontation in the offices of the magazine, raging. Campbell found him at the elevator door.

> He regarded me for a while. I looked back at him, shook my head, sighed, felt myself shaking as the sound of despair oinked out.
> A twinkle came into the Campbell eye as he surveyed it all. "Don't worry about it, son," he said judiciously. And kindly after a little pause, "I just like to shake 'em up" [Malzberg. 1982, pp. 75–76].

Ψ

Exploration of fascinating ideas and possibilities, formulated as propaganda! Science fiction had seen this already, to a degree, in Campbell's unfortunate backing for Hubbard's Dianetics, before he lost faith in that "modern science of mental health" and its trifling hypotheses and simplistic models of mental functioning. His advocacy of psionics was more fruitful and persistent, although it never managed to convince the world that psychic forces were on the verge of being understood and applied.

Ironically, when that did finally happen (in the Star Gate program of operational military remote viewing), the news was heavily classified for nearly two decades at the top secret level before finally being released to the public in 1995 when the program was closed by CIA with a clear and misleading implication that psi didn't exist, bad luck, so sad. (Close reading of the public dismissal of psi shows that no such claim was made; rather, that psi was not reliable enough *to stand by itself* as a crucial information source. But that is true of *every* highly secured data source. Something else was going on. Even today, nobody outside the highest levels has any certain idea what that might have been.)

By then, indeed even before that program was funded and launched, Campbell was dead at the shockingly early age of 61, in 1971. (His first great find, Robert A. Heinlein, seriously ill as that former TB patient was, much of the time, would not die until 1988, his last book released the previous year for his 80th birthday.) As we'll see, Campbell's influence continued—sometimes at second and third hand, from protégés who were too young to know where these notions had been propounded most forcefully in fiction.

In psience fiction, in fact.

Ψ

Professor Peter Lowentrout has argued interestingly (1989) that this permeation of science fiction by psi took four distinct paths, "three of which have in their assumptions concerning psi paralleled the three most common theoretical understandings of it." The first was the "wild talents" tale (not to be confused with the naturalistic approach of Wilson Tucker's 1954 novel *Wild Talents*), akin to the development in comics and some current sf and horror television and movies where anything goes, the splashier the better. As critic and sf writer Paul Di Filippo (2012) has noted, "Ever since the birth of Superman in 1938, the editors, writers and artists of comics have been swept up in a manic, competitive quest to invent and elaborate every conceivable extraordinary ability under this or any other sun."

The second is psi as "biological radio," as in the tendriled slans of A. E. van Vogt's classic sf novel *Slan*, where mutant transhumans possess telepathy due to "growths from formerly little known formations at the top of the brain" (*Slan*, p. 68). This is what I refer to below as treating telepathy and other modifications of communication as if cell phone circuitry had been implanted into the brain. Sometimes this is rendered on the page as if it were an ordinary conversational channel, sometimes as a babble of voices superposed, sometimes as sheer painful gibberish from the unconscious of passers-by; all of these narrative options are exampled below.

The third is what Lowentrout called "the 'metaphysical' SF psi story," which he sees as a profoundly religious or at any rate ontologically deep account of humanity's hunger for transcendence. This is difficult terrain, to which we shall return toward the end of this book.

The fourth approach to psi uses it as a simple plot device, although unlike with the "wild talents" variety the writer has to take pains to reconcile these purported abilities and their consequences with what is already established as accredited science. As with canonical physics, chemistry, biology, geography, and the rest, accreditation can be lost or gained with the arrival of new facts and fresh theories, which provides a small escape chute for plot ideas that seem in any given epoch adamantly beyond the limits of known science. After all, species evolution and plate tectonics were once regarded as heretical, while the luminiferous ether was assumed to be self-evidently real and utterly consequential. This historical instability can offer limited permission to grab a crazy idea (faster than light space travel, time machines, psi) and take it for walk.

Psience fiction has appeared in all four of these categories, very often following the fourth approach. But even in this sub-mode, which seems to capture the essence of what editor Campbell encouraged in his stable of increasingly expert story-tellers, the hidden metaphysics of psi as a door into the transcendent often enriched the nuts-and-bolts engineering technocratic cast of *Astounding*. This element had always been part of science fiction, of course, as can be plainly seen in novels such as Olaf Stapledon's *Odd John* and *Starmaker*.

With this analytic apparatus tucked under one arm, let us look now more closely at fifty or so examples of psience fiction, from standalone novels to series and back to the short stories and novellas where many of these crazy ideas found their early experimental application.

1.
Donald Macpherson (George Humphrey)
Go Home, Unicorn (1935)

Scholars have offered various dates and identities for "the earliest science fiction," sometimes reaching back to the magical devices of ancient mythology but often these days beginning with Mary Shelley's *Frankenstein: The New Prometheus* (1818), and getting its name in 1926 with the launch of Hugo Gernsback's *Amazing Stories* in 1928. Granted, the inventor from Luxembourg called it "scientifiction," an unwieldy blend of "scientific" and "fiction," but the more explicit term "science fiction" soon developed. The heyday of psience fiction was the 1950s, but it had its generic infancy at the start of the 1940s and, as noted, grew apace in the 1950s under John Campbell's influence.

This does not mean that fictional use of the paranormal was never indulged prior to Campbell's golden age. As we have seen, Olaf Stapledon's *Last and First Men* proposed that future species arisen from the human stock would routinely communicate by telepathy. The influence of spiritualism, Rosicrucianism and Theosophy pervaded the arts at the turn of the 1900s like a haze of ectoplasm. One example from 1935, Donald Macpherson's *Go Home, Unicorn* (and its 1937 sequel, *Men Are Like Animals*, not discussed here) serves to exemplify an uneasy blend of psychology, physics and the paranormal in a mode that might be dubbed "between jest and earnest." That quaint and rather fusty phrase was in fact appended less appropriately to another 1935 volume, Stapledon's *Odd John*, which we shall examine in the next chapter.

"Macpherson" was the *nom de plume* of polymathic British psychologist George William Humphrey (1889–1966), a student in Leipzig under the experimentalist Wilhelm Wundt, with a doctorate from Harvard, and later elected a fellow of the Royal Society of Chemistry. His literary character Reggie Brooks, a British professor in Montreal, Canada (as Humphrey was in Queen's University, Kingston, Ontario), invents a new kind of X-ray generator and explores its mutational effects on the genetics of guinea pigs. A series of bizarrely impossible and perhaps supernatural events capture the attention of Brooks, his beautiful upper middle class fiancé Olive Paynter, the raffish journalist McTavish, and Olive's sweet rival Mary Raiche.

A severed hand flies through the open window of a car, buzzing the driver and causing a crash. A severed woman's head with a dreadful face manifests at a dinner party and menaces Mary with "a glance of burning intensity, of bitter and passionate malevolence, a look which seemed to contain all the hatred of woman for woman since the world began" (p. 118). Mary has already been pinched painfully by invisible sharp fingernails. A celebrated composer, conducting his own appalling *Hymn of Hate* that arouses fury in the audience, is flung across the stage by "a mass of shadow" to "fall in the right gangway of the theatre" (p. 77). Other oddities ensue, as Brooks and his coterie attempt to find a rational explanation and put an end to these paranormal assaults.

The novel is a curious blend of social comedy, romance and largely unconscious sexual intrigue, Freudian displacements, and nonsense science that leads to a psience fictional explanation of these various paranormal intrusions. The mutational science is nonsense with no blame to Humphrey, since he was writing two decades before Crick and Watson untangled the secret of DNA's structure and the way its genetic code operates in building proteins, shaping a developing zygote and generally running the body and brain. Reggie's secret ingredient, an X-ray device that sprays the guinea pigs for hours at a time, however gently, would have served in reality only to smash more and more codons into junk like a stream of bullets ripping through the cells.

Perhaps Humphrey suspected something like this might be entailed, because Brooks finds that more and more of his test subjects are ailing and dying. For his explanation of all these connected miseries, however, he turns to proposals by Julian Huxley and other great scientists of the day. Versions of these notions seem to recur in one of the central tropes of psience fiction: the power of Mendelian mutation to bring about not just slow, incremental Darwinian modification, but vast leaps of coordinated change, sometimes "hopeful monster" improvements such as telepathic or clairvoyant powers. In this novel, however, Humphrey drew upon a notion that still recurs in real parapsychological research—the idea of a mind field, directed by deliberate or unconscious intention, midway in nature and effect between mind and matter.

His citations are genuine, although often skewed from their intended meaning. William James and Aron Gurwitsch are noted, although the latter's *The Field of Consciousness* really relates to the phenomenological state (Gurwitsch studied with Husserl) of being aware of things, rather than something like a magnetic or electric field. Kurt Lewin (1890–1947), an important German social psychologist, is quoted on "the psychic field" (p. 94), and

Julian Huxley, former Professor of Biology at London University, has stated that science has established the existence of a substance that is between thought and matter; you may also remember that his statement was corroborated by Fraser-Harris, the well-know Professor of Neurology at the Medical School at Halifax. You may further recall that the existence of this half-way substance was certified not so many years ago by twenty-six professors of German universities [p. 95].

Mary suggests calling this stuff "electroplasm" (rather that the spiritualist term "ectoplasm" or "teleplasm") and Brooks accepts the coinage. As the adventure continues, they learn that bright light can cause objects forged by this uncanny "mind-matter field" (p. 99) to shrink and rot, sparing the heroes from various lethal depredations. After Reggie and Olive attend a lecture on mythological creatures (no *Game of Thrones* to while away the hours in the mid-1930s), Brooks's experiments elicit the titular unicorn. The size of a pony, it is even so a ferocious creature, not at all the gentle white creature of children's fairy tales. Olive, locked in a room with the feral brute, saves herself only by clambering into a closet. Brooks is almost stabbed fatally. Only when young Mary enters the room, filled with confidence, does the mind-matter composite animal change its tune. "The beast, so furious a moment ago, was standing by, fawning up at her.... In amazement, they watched her stroke its noble head, its milk-white flanks" (p. 209). As any reader of fantasy will know at once, this event is a cause of deep embarrassment to Mary, and of irritation to Olive; unicorns are tamed only by virgins. Olive, publicly unmasked for her previously unacknowledged "indiscretion," quickly returns her expensive engagement ring to Brooks and heads for the Himalayas, shooting a tigress *en route* and sending home the skin.

Leaving aside the sexism of the period, how were these miracles induced? The guinea pigs dunnit! Mutated by the flood of X-rays, they collectively generated a sort of morphic field that spread from the research lab into the nearby environment, open to modification by any strong emotional impulses. Brooks provides clarification, quoting the Berlin professor of chemistry and Nobel laureate Wilhelm Oswald (1853-1932): "Certain persons have the power of transforming their store of physiological energy ... into other forms, which they can project through space and transform into one of the known forms of energy at given points in space" (pp. 229-30). Julian Huxley is invoked again:

Here we are in the presence of a new phenomenon, a new mode of interaction between the realm of mind and the realm of matter, a new way in which thought and will can be translated into practical action.... It may be possible for many persons to produce it simultaneously, and so, perhaps, to multiply its intensity and its range of action manyfold, as we can multiply electric power by linking a number of batteries [pp. 234-35].

Years later, this conjecture was expressed vividly in a short psience fiction story, "The Public Hating," a bitter allegorical story of America's McCarthyist period, published in 1955 by the acerbic satirist Steve Allen. "The Public Hating" gave a fictitious but frightening account of mob persecution of minorities via mass psychokinesis:

> I think it was that guy at Duke University first came up with the idea. The mind over matter thing had been around for a long time, of course…. Then one time they got the idea of taking the dice into an auditorium and having about 2000 people concentrate on forcing the dice one way or another. That did it. It was the most natural thing in the world when you think about it. If one horse can pull a heavy load so far and so fast it figures that 10 horses can pull it a lot farther and a lot faster. They had those dice fallin' where they wanted 'em 80 percent of the time.[1]

The assembled crowd, with their hatred coordinated and magnified by their leader, effectively burn a political criminal to death by sheer willpower. Allen's fictional narrator, and Huxley, were right—it *is* a natural idea to combine the psychic intentions of numerous people for tests of precognition and telepathy as well as psychokinesis. It turns out not to be so easy in the real world, perhaps fortunately. (The Global Consciousness Project, run for many years by Dr. Roger Nelson and still going, looks for similarities in the chance deviations from average, in a number of random event detectors scattered around the world, when pre-specified events of major significance occur— the 9/11 mass murders, for example. Nelson finds many such events that correlate with global shock or excitement but not at other neutral times. His colleague Peter Bancel, in a *tour de force* of nuanced analysis, showed that this result is likely to be a psychokinetic effect of Nelson's own intentions.)[2]

Even more frightening than Steve Allen's imaginary story of murderous mass hatred, or the B-movie final horror passages in *Go Home, Unicorn*, is the story reported of the author in the *Queen's University Encyclopedia*: "George Humphrey was extremely interested in human nature. There is a story that when his daughter was born, he held her out the second story window 'to see if she would register fear.'"[3]

2.
Olaf Stapledon
Odd John (1935)

When the British Communist yet cryptically religious philosopher William Olaf Stapledon wrote his handful of great proto-sf novels, he was less interested in producing reflective portraits of upper-middle class life, after the Bloomsbury model, and more driven by a wish to allegorize the cosmic vector of intelligence, from the individual mind at its highest pitch (even that of a supremely gifted dog) to the bleak rise and fall into nothingness of the entire universe.

So when he chose for *Odd John*, his novel of mutant genius, the clumsy subtitle *A Story Between Jest and Earnest*, he did not mean that blend of arch whimsy and mock psychophysics abundant in "Donald Macpherson's" novels *Go Home, Unicorn* and *Men Are Like Animals*; he wanted to convey something like "playfully inventive but intended seriously." The mysterious and sometimes abominable workings of John Wainwright's supercharged mind were not offered as an *amusette*, for idle pleasure, but as a thought experiment in moral choice and imaginative but constrained extrapolation—taking evolution seriously, casting forward from real world "freaks of nature" to beneficial gene mutations that would yield the species after humanity: Homo superior.

Would this, additionally, be at last a recipe for a workable utopia? No, because these young monsters would be forced by the ambition and dread of imperial colonialists to destroy their island home in what amounts to a self-inflicted nuclear explosion.

The historian and science fiction critic Edward James offered cautionary remarks in his chapter "Utopias and anti-utopias" in *The Cambridge Companion to Science Fiction* (2003), drawing on elements of psience fiction:

> There are numerous themes of modern sf which should probably be regarded as part of the utopian project. Back in the 1950s, for instance, there was serious scientific work on telepathy and other forms of ESP, most famously by J.B. Rhine at Duke University, and therefore speculation along these lines was justified as science and not fantasy. What kind of society might emerge if all thoughts were open to all and perfect harmony and understanding—the goal of utopian writers since More—could be achieved? ... When one imagines such changes, it is possible to think of just as many dystopian outcomes as utopian ones. Universal telepathy might bring mental harmony; it might bring political control and the end of privacy.

Professor James adds, illuminating Stapledon's own quest: "The unasked but essential question in most utopian novels—'what is the meaning of life?' or 'what is the destiny of man?'—is a question raised by almost no one these days apart from theologians and sf writers. It is the ultimate, unanswerable, question" (p. 228). These are questions raised forcefully, if answered enigmatically, in this remarkable novel.

Ψ

The book's adult narrator, or historian, is initially a bored, uninterested journalist, and later the extremely wealthy custodian of young John's brilliant inventions and investments. His markedly "*Hom. Sap.*" inferior position is registered by John's fond nickname for him: "Fido." His initial attraction to the Wainwright family is not the rather gruesome and apparently retarded baby John, nor the physician father, but the child's mother Pax, described in blunt terms as "a great sluggish blonde" of " extraordinary dumbness" (p. 18). Pax loves her mutant son unreservedly—and indeed is the one who names him "Odd John"—and in the adolescent boy's grimmest days of loneliness and dispossession comforts him sexually. The narrator, in the 1930s, cannot bring himself to speak directly of this incestuous relationship, but neither does he moralize about it, nor condemn his young master for several murders, agreeing that all these improprieties are simply necessary, however regrettable.

John is not recognized as a "supernormal" genius for years, after a gestation of eleven months and eventual birth, in appearance born two months before term. The development of this intriguing and disturbing oddity is traced with remarkable devoted objectivity, citing explicitly the earlier wunderkind novel by J.D. Beresford, *The Hampdenshire Wonder* (1911) but taking that tale of Victor Stott as literally true.

John is clearly a mutation or sport, and his task becomes the discovery and gathering to utopian safety of others of his kind, the "wide-awakes," some of them hideously deformed, others mentally ill from their treatment by humans, a few quite conventional in appearance but all with a range of preternatural gifts. The book is more than halfway through, however, before it becomes an important early instance of psience fiction, as this search for other supernormals draws upon telepathic communication between the widely separated mutants. Most ingeniously, this mental link proves to have a kind of precognitive or postcognitive aspect. John finds himself in contact with an itinerant sage, Adlan, in Port Saïd who turns out to have died thirty-five years earlier. Born in 1512, he was very old and quite near death when John contacted him, reaching back to 1896.

A boatman, Adlan was utterly unconcerned with pretension or physical comfort, sending his mind back into the deep past, exploring the great times of Egypt and then other places via a blend of focused intellect and telepathy. "[B]y conjuring up all that he knew of John Ruskin ... he could make contact with that didactic sage" who had died in 1900. (This device will reappear in Colin Wilson's psience fiction novel *The Philosopher's Stone*.)

Most importantly for atheistic John's development, Adlan was a Muslim mystic. Soon, in the past, Adlan dies, restful, worshipful, but unprepared to praise the horrors of the future. Telepathy remains John's principle mechanism of finding and gathering others of his kind. These supernormals are never genetically related but presumably brought into being by some kind of Bergsonian life force.

ESP is not the only paranormal ability John uncovers and masters. His machines are powered by a form of what seems to be sub-nuclear fusion produced by suspension of symmetry-breaking, activated and controlled by a kind of psychological enticement of matter. (A similar method was postulated nine years later by A.E. van Vogt in "Far Centaurus," the "adelidicnander force" governed by the psychology of electrons, which are open to persuasion.[1]) How is this done? Stapledon indulges both his ingenuity and whimsy, delightfully:

> "it's not a physical key at all but a psychical one. It's no use trying to *overcome* those terrific interlocking forces. You must just *abolish* them for the time being; send them to sleep, so to speak.... What I do, then, is to hypnotize the little devils so that they go limp for a moment and loosen their grip on one another. Then when they wake up they barge about in hilarious freedom, and all you have to do is to see that their barging drives your machinery" [p. 119].

John explains with a certain frustration that electrons and protons are "not *really* independent entities at all, but determinations within a system— the cosmos. And they're not *really* just physical, but determinations within a psychophysical system" (p. 120).

Essentially, then, the psience fictional aspects of this notable novel yield a unified field theory of matter and energy that is manipulable by intention and will. This is a proposal that might make more sense today than it did when Stapledon was writing, and is close to the metaphysics underlying some schools of parapsychology. Like the mind-matter fields in Macpherson's novels discussed in the previous chapter, and perhaps the Global Consciousness field that allegedly transduces individual grief or joy on the largest scale into "disruptions of the Force" (as *Star Wars* would have it), the substrate of *Odd John*'s universe seems at once vitalist and quantum mechanical.

The vector of this strange novel, like that of Stapledon's own life, is toward worship ... but of what pantheistic deity we are never certain. Under the long-distance telepathic guidance of a blind Tibetan monk.

In the end, despite the ruinous intervention of naval vessels from the Six Great Powers, the supernormals tidy their utopian gardens and homes, embrace each other, and take their terrible quietus. "Suddenly there was blinding light and noise and pain, then nothing" (p. 190). Fido, journalist and faithful handservant to a species that presumably still exists in small isolated individuals scattered through the globe, fulfills his promise to write John's biography. He is a husband and father, and by now has been long dead himself, like Olaf Stapledon. But his extraordinary work of psience fiction persists, to our astonishment and enrichment.

3.
E. E. Smith,
The History of Civilization (1937/1951)

In some respects the very founding moment or document of space operatic science fiction in English was E. E. "Doc" Smith's "Lensmen" sequence, *The History of Civilization*. It began in 1937–38, despite some later retrofitting, with hero Kim Kinnison's admission in *Galactic Patrol* (serialized in the pre-Campbell *Astounding*, revised 1950) to the gloriously virtuous ranks of the select human and nonhuman guardians of the nascent Civilization of the Milky Way. This saga was conducted under the hidden guidance of Mentor, one of the ancient Arisians, in a multi-billion year war against the extra-dimensional evil of Boskone/Eddore. This breathless adventure accelerated from volume to volume, propelled by rapidly advancing science and military technology and the emergence of heroes via a long-developed eugenics breeding program to create the perfect warriors in this angelic Arisian cause. (Several decades later an equivalent dynamic would power Frank Herbert's far more sophisticated *Dune* sequence, with the prophetic Paul Atreides in the upgraded role of Kinnison *pere*.)

The second and arguably the most important volume was *Gray Lensman* (serialized 1939–40, revised 1951), followed by *Second-Stage Lensman* and the final belated volume (1947–48, revised 1954), *Children of the Lens* (akin to Herbert's *Children of Dune*, but more so). The logic of the saga culminated beyond the last page in a sort of (then unpublishable, and hence unwritten) necessary transition that we can see must have been the incestuous mating of Kinnison's and his wife Clarissa's prodigy son Kit and their equally magnificent pair of twin daughters: Kathryn and Karen, and Camilla and Constance.

Smith's writing is often excruciatingly bad. Here is a random example:

"And so on, for one solid hour!" Jill snarled, as she snapped the switch viciously. "How do you like *them* potatoes?"

"Hell's—Blazing—Pinnacles!" This from Jack, silent for seconds, and:

"Rugged stuff ... very, very rugged," from Northrop. "No wonder you look sort of pooped, Spud. Being Chief Bodyguard must have developed recently into quite a chore."

"You ain't just snapping your choppers, bub," was Costigan's grimly flippant reply. "I've yelled for help—in force" [*First Lensman*, p. 220].

And yet the sequence as a whole can be propulsively exciting, for those who can tolerate its pulp vulgarity. Here is another example, from *Second-Stage Lensman*:

The mental blast came ahead even of the first word, but the Gray Lensman, supremely ready, was already in action. One quick thrust of his chin flicked off the thought-screen. The shielded cigarette-case flew open, his more-than-half-alive Lens blazed again upon his massive wrist. His blaster leaped out of its scabbard, flaming destruction as it came—a ravening tongue of incandescent fury which licked out of existence in the twinkling of an eye the Bergenholms' control panels and the operators clustered before it. The vessel went inert—much work would have to be done before the Boskonian flagship could again fly free!

These matters required only a fraction of a second. Well indeed it was that they did not take longer, for the ever-mounting fury of the prime minister's attack soon necessitated more—much more—than an automatic block, how ever capable. But Kimball Kinnison, Gray Lensman, Lensman of Lensmen had more—ever so much more—than that! [*Astounding*, vol. 28, no. 6, 1942, p. 109].

Despite such abundant and barely tolerable weaknesses, critic John Clute has declared the series, "one of the greatest stories ever told ... a tale that expands remorselessly in space and time."[1] Elsewhere, he comments:

As the central sequence progresses, we climb with Kinnison, link by link, a vast chain of command, until he defeats Boskone and becomes, in essence, the ruler of the civilized universe. But Kinnison is destined only slowly to understand that the final empire of Boskone he has destroyed, through the use of weapons of unparalleled immensity, is not the final enemy, whose name he never learns, no more than he ever discovers the full truth about his own Arisian mentors, whose civilized precepts he enacts and whose apparent true form ... is a further layer of disguise. His powers are vast, though he remains ignorant of the true scale of the Universe, which is greater, and requires greater powers to confront, than even a Hero with Superpowers is capable of grasping.... Today, while he must be read by anyone interested in understanding the deep appeal of American Genre SF in the days before World War Two, any revisit to his work should be made in the loving awareness that he is a creature of the dawn.[2]

The salience for psience fiction of this grand if crudely wrought saga is its central motif: the Lens, specially rendered for each Galactic Patrol graduate of sufficient worth, bravery, purity of motive, etc. This device is less a machine than a psi enhancer, delivering telepathic and perceptive powers to its wearer and perishing with the bearer's death. It is, crucially and mysteriously, "not essentially scientific in nature," Kinnison is told. "It is almost entirely philosophical, and was developed for us by the Arisians" (*Galactic Patrol*, p. 9).

Remarkably, the "us" mentioned are not just heroic and almost entirely male humans but members of other and very non-human interstellar species as well. Unlike many of the pulp fantasias of its earliest epoch, the Lensman saga is surprisingly unbigoted, despite its sentimental sexism and even there Smith was refreshingly open to the idea that women were autonomous and intelligent. One of them, Kinnison's fire-haired wife, Clarissa, is even accepted as the Red Lensman, and finally undergoes the rigors of that discipline's Second Stage.

While no material Lensman (or, perhaps better, *Lenticulares sapiens*) could understand the psi machine's operation, Kinnison was told that

> "thinking of your Lens as being synchronized with, or in exact resonance with, your own vital principle or ego will give you a rough idea of it. The Lens is not really alive, as we understand the term. It is, however, endowed with a sort of pseudo-life, by virtue of which it gives off its strong, characteristically changing light as long as it is in metal-to-flesh circuit with the living mentality for which it was designed. Also by virtue of that pseudo-life, it acts as a telepath through which you may converse with other intelligences, even though they may possess no organs of speech or of hearing" [p. 10].

In short, the Lens is a very superior version of what John Campbell would latter dub a "psionics machine" and others would refer to as a "radionics device": an amplifier and transducer of the mind, enabling psychic resonance between two or more brains—not just "thought transference" as it had been known by mediums and others but a sort of interpenetration of experience. Facilitated by an alien Lensman that is "a thirty-foot long, crocodile-headed, leather-winged python" (p. 84), Kinnison's human mind is placed in rapport with just such an extraterrestrial dragon:

> Kinnison relaxed his mind completely, and that of the Velantian came welling in; wave upon friendly, surging wave of benevolent power. And not only—or not precisely—power. It was more than power; it was a dynamic poignancy, a vibrant penetrance, a depth and clarity of perception that Kinnison in his most cogent moments had never dreamed a possibility. The possessor of that mind knew things, cameo-clear in microscopic detail, which the keenest minds of Earth could perceive only as chaotically indistinct masses of mental light and shade, of no recognizable pattern whatever! [p. 86].

Smith here conjectured a form of mental fusion that the 1960s TV show *Star Trek* borrowed as the Vulcan "mind meld." But what of ordinary uses of the Lens as a telepathy augment? Young Kinnison explains to his Valerian (or heavy-gravity) companion vanBirskirk, after an attack by unknown space pirates, how he can understand their jargon:

> "The Len receives as pure thought any pattern of force which represents, or is in any way connected with, thought. My brain receives this thought in English, since that is my native language. At the same time my ears are practically out of circuit, so that I actually hear the English language instead of whatever noise is being made. I do not hear the foreign sounds at all. Therefore I haven't the slightest idea what the pirates' language sounds like ...
> "Conversely, when I want to talk to someone who doesn't know any language

The History of Civilization (1937/1951)   31

I do, I simply think into the Lens and direct its force at him, and he thinks I am talking to him in his own mother tongue" [p. 40].

There are clearly some confusions here. Kinnison does not engage in telepathy per se, since the unknown language must first be intercepted and translated by his Lens. Yet his answers, translated in turn into the unknown language, are somehow imposed on the listener despite his or her lack of an intermediary Lens. Kinnison adds, "Of course I can broadcast a thought— everybody does, more or less—but without a Lens at the other end I can't reach very far" (p. 41). Curiously, this account is followed by what seems a recognizable portrait of today's practice of remote viewing, or more exactly of clairvoyance facilitated by some sort of telepathic exchange:

> "You can receive a thought ... everybody broadcasts.... Then you can read minds?" ...
> "When I want to, yes. That's what I was doing while we were mopping up. I demanded the location of their base from every one of them alive, but none of them knew it. I got a lot of pictures and descriptions of the buildings, layout, arrangements and personnel of the base, but not a hint of where it is in space" [p. 41].

We might compare this with a description, now declassified, by U.S. military remote viewer Joe McMoneagle. His monitor, who is blind to the target McMoneagle is meant to discover and describe, asks him after receiving earlier details:

> MONITOR: Tell me more about this room that you're in.
> McMONEAGLE: There's a ... it's got a high, high ceiling and there's a ... like compartment rooms passenger's side. Like a ... a look like a ... glass enclosures ... some sort. I feel very small in this room ... uh
> McMONEAGLE: I see cranes ... for some reason ... uh ... lift devices, support on girders or something.
> MONITOR: Go up to where these lift devices are on these girders, and look down. Go up and then look down and describe what you see.
> McMONEAGLE: Looks like a.... I'm looking down like I was standing in the bottom of this ... uh ... like a ramp type thing ... a wide ramp. [...] the walls are high on the ramp where it comes in but they narrow down towards the ... uh ... center of the floor ... toward the center of the room. Feels like a ... some kind of shelves ... shelves are ... uh ... cat walks or something on the right [McMoneagle, in *Extrasensory Perception*, vol. 1, 2015, p. 292].

This kind of agonizing extraction of detail continues for several hours, and builds into a truly remarkable and accurate image of what was at the time an unknown Soviet submarine under construction. In the real world, then, the psience fictional powers of the Lensmen make "Doc" Smith's imaginary scenario seem almost ... precognitive.

4.
A. E. van Vogt, *Slan* (1940/1946/1968)

The canonical Golden Age novel of telepathy, by *Astounding*'s then-recent major discovery, Canadian Alfred Elton van Vogt (1912–2000), is a tangled yet drastically simplified tale of political chicanery in a prejudicial future war against mind-reading mutants, the slans. It is set at least 1500 years from now.

Slans have two hearts, for extra endurance, turbo-charged musculature, enhanced genius—an adult slan's "intelligence is two to three hundred percent higher than that of a normal human being" (chapter 4, p. 20). Golden thread-like tendrils mediate their telepathic abilities, rising from their hair and waving slightly as if in a breeze.[1] Are slans even human any longer? The baseline humans hate and fear them, convinced that they begin as stolen and then mutilated human babies. Rewards are posted for the death of any slan careless enough to be outnumbered by a mob or trapped by the secret police. Here is the first and immediate glimpse we have of slans on the run, handled quite deftly:

> His mother's hand felt cold, clutching his.
> Her fear as they walked hurriedly along the street was a quiet, swift pulsation that throbbed from her mind to his. A hundred other thoughts beat against his mind, from the crowds that swarmed by on either side, and from inside the buildings they passed. But only his mother's thoughts were clear and coherent—and afraid.
> "They're following us, Jommy," her brain telegraphed. "They're not sure, but they suspect. We've risked once too often coming into the capital, though I did hope that this time I could show you the old slan way of getting into the catacombs, where your father's secret is hidden. Jommy, if the worst happens, you know what to do. We've practiced it often enough. And, Jommy, don't be afraid, don't get excited. You may be only nine years old, but you're as intelligent as any fifteen-year-old human being."
> Don't be afraid. Easy to advise, Jommy thought, and hid the thought from her. She wouldn't like that concealment, that distorting shield between them. But there were thoughts that had to be kept back. She mustn't know he was afraid also [opening of chapter 1, p. 5].

Here we have in *précis* an image of ESP—telepathy, in this case—that would endure in a great many subsequent psience fiction stories. Little Jommy, far from mature as a telepath, experiences two imbricated or enfolded species of ESP: the "telegraphic" or better "telephonic" exchange of sentences, as if psi were a cellphone implanted in the brain; plus a generalized hubbub of "mind clamor" from the surrounding crowds. We see that Jommy and his mother Patricia are not bombarded into overload and incapacity by all the thoughts and emotions of the whole world, so there are limits to the range and efficacy of slannish ESP.

In the real world of disciplined paranormal research, is psi actually anything like this? Mind-chat, strings of linear words and sentences, seems rarely reported, except during mediumship. Still, something *non*-psychic like this can occur during moments of drifting into sleep (hypnagogia). It is less often observed as one wakes from a dream (hypnopompia). The hiss of a white noise generator can be seized by the brain in this threshold or liminal state and misinterpreted as a voice, perhaps speaking one's name or muttering a few words.

This is similar to the "cocktail party" effect experienced in a room filled with chatter through which a mention of your name or a particularly obsessional interest "pops out," intercepted and clarified by trained feature detectors in specialized regions of your brain. In the hypnogogic case, however, there is no real cocktail utterance to snag your attention and turn up the gain, but those same feature detectors are rather good at alerting you to what seems like a communication relevant to your world but is really just masquerading as such. (This neurological accident is known as Pareidolia, which also explains why that random burn on a piece of toast, or shadow cast by a tree, seems to be a compelling and supernatural portrait of Jesus or Shiva.)

A. E. van Vogt plays changes on this telepathy theme as the novel progresses. But it is important to point out at once that *Slan* is not centrally a drama of psi, but rather, like Stapledon's *Odd John*, of a transcendental superman genius, *Homo superior*. The novel was van Vogt's clever reply to John Campbell's dismissive assertion that it was impossible to write successful fiction about a superman/ superwoman because we humans with a mean IQ of 100 and almost no examples of any with IQ 200 are therefore prohibited from imagining with any accuracy what mutants with an IQ between 200 and 600 could be like, how they would think and resolve problems, their emotional reactions. Yet all of these unknowable features are necessary for an emotionally engaging story. Ah ha, replied van Vogt. That might be true of an *adult H. superior*, but what about an unformed, immature child?

Campbell was delighted by this suggestion, and urged van Vogt to write his book. Thus, *Slan*, which *Astounding* published as a fabulously successful four-part serial in 1940. It presents the *Bildungsroman* of two pre-adolescent tendriled slans, Jommy at 9, both his parents murdered, and orphan 11-year-old Kathleen Layton, the experimental charge of (apparently human) Kier Gray, "absolute dictator of the entire planet" (p. 5).

To thicken the plot and the problematic, van Vogt adds a hidden subcategory of slans, tendrilless and hence lacking ESP but perhaps equally brilliant, with technologies far in advance of the human population and with spacecraft and a base on terraformed Mars. These latter are equally eager to rid the world of the telepathic tendriled, whom they consider a terrifyingly dangerous *casus belli*. The furtive, hunted tendriled, whose ancestors were briefly masters of the world, do their best to stir up the tendrilless, toughening them for their future resurgence: mutant vs. mutant. On the other hand, Jommy Cross (the Christian symbolism of JC's name is surely no accident) allegedly wishes to put an end to this three-sided protracted genocide.

Ψ

Much of the narrative, especially with Jommy's emerging maturity, is hyperbolic and enacts van Vogt's own emerging enthusiasm for versions of the "Great Man" (or indeed Great Übermensch) theory of history. Since this study emphasizes the psience fictional aspects of the books under consideration, we can safely leave this crowded narrative and move directly to a summary of what Jommy learns about his people and their history and prospects, which is largely aggregated in *Slan*'s final chapter. An excellent online analysis of the novel, with carefully calculated timeline and background, is Isaac Walwyn's "Slanology," which can be investigated fruitfully both before and after reading van Vogt's own repeatedly revised text.[2]

In the 36th century, history is traumatically garbled. It is widely accepted by ordinary humans that for several hundred years random mutations in human stock afflicted an immense number of generally deleterious pregnancies and births. In the subsequent grief and increasing hysteria, the small proportion of eugenic births are alleged to have been created by a "mutation machine" designed by Samuel Lann. In reality the triplets (one male, two female babies) Lann finds and raises are the first known instances of this new species, dubbed "slans" from his name. Not long after puberty, the twin girls are impregnated by their brother and subsequent generations continue this pattern of multiple offspring per pregnancy. This surge in slan births is attributed by rumor to artificial methods that also (somehow) create the vast number of damaged babies.

Given their spectacular gifts, not least telepathy, a group of slans takes command of the world, but the genocidal War of Disaster leaves only a small number of slans in hiding. For centuries, slans continue to be hunted as they search for ways to modify their new genes, helped along by surgeries that reshape them into simulated normals. These are the tendrilless slans, protected against even advanced scanning methods. Yet Jommy learns that this adjustment is temporary; eventually the telepathic abilities will reemerge, along with the tendrils and enhanced physiology. Meanwhile, the bitter centuries of genetic uproar occasioned by technological stress and pollution will gradually render the classic human genome ever more dysfunctional and finally sterile. The slans hope to ease this species extinction via a tranquilizing "final solution"[3] (p. 154) based on Jommy's invention of "hypnotic crystals" and similar soporifics.

Ψ

Of particular interest to our investigation, though, are van Vogt's representations of psychic awareness, communication, and defense against both the torrent of ambient thoughts and specific bursts of unspoken bigotry and hatred against slans. Kathleen, living as she does in the central redoubt of the normals, suffers so terribly from this toxic noise that she learns to shut off her attention from its endless attack—and hence misses some cues to her imminent murder. Luckily she averts this fate, although, at age 21, midway through the novel, she meets Jommy, they fall rapturous in love, and shockingly Kathleen is slain by John Petty, the venomous and anything-but-petty chief of the secret police.

This crux is presented in Chapter 14, and conveys for the first time since the death of Patricia Cross a full psychic mutuality. Jommy detects a search by human foes for "a slan girl," and drives into the danger zone. Abruptly, the psience fiction premise is extended to linear text from a psi machine:

> an outside thought touched his mind.
> "Attention, slans! This is a Porgrave thought-broadcasting machine. Please turn up the side road half a mile ahead. A further message will be given later" [p. 109].

Jommy enters the garage of a rickety old house, and Kathleen emerges with a dazzling smile.

> And she was a slan!
> And he was a slan!
> ...for Kathleen, who had never had to conceal her thoughts, the surprise was devastating. She fought for control and found herself uncontrollable. The little-used shield was suddenly, briefly, unusable [p. 110].

And here is the critical moment of total mind-sharing:

> There was a noble pride in the rich flow of thought matter that streamed from her mind in that instant when her brain was like an open, unprotected book. Pride, and a golden humility. Humility based on a deep sensitivity, an immense understanding that equaled his own, yet lacked the tempering of unending struggle and danger. There was a warm goodheartedness in her that had nevertheless know resentment and tears, and faced limitless hate.
> ...It was not a thought, but an emotion; all sad, sweet, glorious emotion [pp. 110–11].

Jommy realizes in epiphany:

> What a rich joy it was to be able to entwine your mind with another sympathetic brain so intimately that the two streams of thought seemed one, and question and answer and all discussion included instantly all the subtle overtones that the cold medium of words could never transmit [p. 111].

And within minutes, a second, ruinous epiphany, as she is shot dead by their slan hunter:

> The shot echoes from her mind to his. For a terrible moment of intolerable strain, her mind held off the death that the crashing bullet in her brain had brought.
> "Oh, Jommy, and we could have been so happy. Good-bye, my dearest—"
> ...he followed the life force as it faded in a flash from her mind. The blackout wall of death suddenly barred his mind from that which had been Kathleen's [p. 117].

Ψ

The romance sentimentality of this scene, and the hasty writing (can something fade in a flash?), is less remarkable than this early effective portrayal of mind-to-mind linkage in extremis, in both their instantaneous loving recognition and the frightful sharing of the moment of death. Of course, being the kind of narrative it is, we can be sure that Kathleen will return to life, and indeed her damaged brain tissues are regrown via Martian advances in tendrilless surgery using "rapid tissue-building rays" (p. 138). But for the moment, Jommy's life takes a blow greater even than that inflicted by the murders of his parents.

Eventually, armed with wonderful weapons and devices pioneered by his father and developed to their full lethality and power by his own theory of atomic structure, Jommy confronts Kier Gray and finds that this champion of the normal is himself a slan, and indeed the father, via a tendriled and now tragically dead slan woman, of Kathleen Layton Gray. United again with his resurrected lover, where can Jommy Cross take this abbreviated saga? Van Vogt made some attempts to sketch out a sequel, but it remained incomplete until the hand of Kevin Anderson fell upon it with the same unfortunate effect that devalued all the endless *Dune* prequels and sequels that he has committed with Frank Herbert's son Brian. In *Slan Hunter*, the terrifying John Petty becomes a giggling buffoon.[4] Logic flails about even more hopelessly than van Vogt's at his worst. Well, not his worst; that was reserved for the truly dreadful *Null-A Three*, apparently written as its luckless author was sinking into the Alzheimer's that eventually carried him to his death.

Ψ

The World of Null-A (1945) was the great and irritatingly opaque novel that followed *Slan* and was followed in turn by its sequel, serialized as *The

Players of Null-A (1948) and later heavily revised as *The Pawns of Null-A* (1956). These are two strange sagas about a two-brained superman clone with the power to be reincarnated at the moment of death into a slumbering adult clone body. Thus Kathleen Layton's recovery from murderous attack in *Slan* was subsumed in the subsequent sequence into a more prescient trope, the combined notions of physical cloning and, as it were, spiritual rebirth or serial immortality. Gilbert Gosseyn (Go-Sane) is an heroic if deeply confused *Übermensch*, trained in the mental rigors of real-world General Semantics developed by Count Alfred Korzybski (1879–1950), as fictionalized and puffed up by van Vogt in his ceaseless quest for transcendence of the frailties of human minds and bodies.[5]

This van Vogtian yearning for the transhuman or transcendental condition, arguably a recurrent core element in nearly all varieties of psience fiction, is anatomized brilliantly and extensively by Alexei and Cory Panshin in *The World Beyond the Hill*, subtitled explicitly *Science Fiction and the Quest for Transcendence* (Tarcher, 1989). We shall return to that abiding concern as the key driver in the development of psience fiction in the Conclusion of this volume.

5.
James Blish,
Jack of Eagles (1949/1952)

Blish's *Jack of Eagles* (1952), an extended version of his novella "Let the Finder Beware" (*Thrilling Wonder Stories*, December 1949), is perhaps the first pure psience fiction novel, or at any rate the first elaborately intelligent one. For all that, it is also annoyingly silly in entirely unnecessary ways. It pads the lean novella with at least one ridiculous and implausible oaf who adds nothing but some gratuitous bumbling action. Protagonist Danny Caiden's uninspiring work-setting, writing copy for the packaging section of a New York "frosted food" industry trade paper, seems to be borrowed directly from Blish's own employment when he was writing the novella. For a time, Blish edited *Frosted Food Field*, and was followed in that post by science fiction historian/critic Sam Moskowitz (1920–1997), who notes "in those days, frozen foods were called frosted foods, to dispel the impression that frozen foods were spoiled foods." Moskowitz later went on "to a much more enticing offer from *Quick Frozen Foods*, the leading journal in the field."[1] Science fiction writers, then as now, had to find some sort of assured income, as did future psi master Danny Caiden.

(One of the minor characters, incidentally, mocks Caiden's name, pretending it's a tribute to the then-famous antic film comedian Danny Kaye. "Your mother must of hung too much around the movies." Caiden is scathing: "The name's a corruption of a New Orleans term that would be familiar to anybody with two brain cells." Presumably he means a corruption, akin to "Cajun," of "Acadian," from the 17th century French settlers in that region. This icy rebuke has no further bearing on the plot or background; it is James Blish expressing his detestation of ignorant oafs.)

The plot and cast of characters seem to have been compiled on the fly, one more or less random thing after another, but with some quite wonderful set pieces popping up but leading to a confused and indeed unintelligible finale. It is the set pieces, often constructions of intellectual architecture, that make this such an important contribution to the developing megatext archive and encyclopedia we are calling "psience fiction." As Blish's friend and sometime collaborator Robert A.W. Lowndes observes in an introduction to the novel, "It was ... the first time in science-fiction that extra-sensory perception (ESP) and psychic phenomena were examined on a scientific, rather than an occult, supernatural, or mystic basis."[2]

The plot ranges from precognition to telepathy, to an ancillary psionic machine that works by modifying one or more of the fundamental parameters of physical quantum reality, allowing teleportation, to a thrilling *noir*-ish chase through parallel universes, what we now call the *multiverse* but which had not yet been invented by scientists in the late 1940s. The key, again, is Blish's educated approach (he had a science degree from Rutgers, majoring in shoreline biology) to these pulpish wonders. Science fiction writer and historian Brian Stableford comments astutely[3]:

> As he gleans information piece by piece, [Danny's] understanding of his powers increases, and as he gains this understanding, his conscious control of them becomes steadily more secure.
> In attempting to comprehend the physics of the various phenomena with which he is dealing, Danny makes oblique references to Planck's constant and an essay by Haldane on non–Einsteinian relativity, and discovers Heisenberg's Uncertainty Principle lurking in the pages of Korzybski's *Science and Sanity*.... [Blish] is committed to the point of view which insists that if the phenomena are real then the scientific method must be adequate to the problem of systematizing them and revealing the techniques of their manipulation. It is always possible, in the literary cosmos to which Blish's fiction belongs, to gain knowledge *and apply it* [pp. 11–12].

This is, in essence, psience fiction's central project: rigorous understanding of psi, then suitable methods of application—although such generic tales have often regressed to magical hand-waving.

Ψ

Danny Caiden's background is not clearly charted, but he has a degree and served in the military during World War II, before undergoing an excessively elaborate Veterans Administration vocational testing that placed him in food industry journalism under the GI Bill several years before the novel opens. It is tempting to map this history against Blish's own: born May 1921, drafted at the end of college by the Army where he served as a laboratory medical technician but was discharged (without court martial) "when he refused an order to clean the grease trap under the kitchen sink."[4]

So Danny would be about 28 or 30 in both the novella and expanded novel. He has no wife or girlfriend, nor, apparently, family. He might be "abnormally normal" but he has a curious minor gift for finding lost objects on request (a kind of low level clairvoyance) and he is shaken by a recent episode of precognition, foreseeing a horrific street crash and deadly pyre. Another premonition gets him fired—he writes up the imminent price-fixing indictment of International Wheat Corp. before any public hint of this is known—and then harassed by the authorities on suspicion of what amounts to insider trading when his immediate stock speculation returns him significant funds.

From the outset, the plot is a cat's cradle of unexplained connections that only start to make sense, to the extent that they do, some two-thirds of the way into the novel. At that point Danny thinks in despair: "Was everyone in this game carrying all his cards up his sleeve?" (p. 171). Unpacking the diegesis is possible, but only in retrospect. The trouble begins right at the start, before we see Danny at work as a lowly journalist for Delta Publishing. His premonition of the fatal street crash has him in mental turmoil, and he detects "soundless" subterranean voices conversing in penny dreadful menacing phrases like a cult ritual:

> "There has been a Decision."
> "Yes. I have estanned the tension...."
> "So many paths—may we never interfere?"
> "No, my brother. All go to the same goal."
> "Let the finder beware" [p. 18].

At the *Food Chronicler* office, Danny is berated by the chief editor for his unsubstantiated price-fixing story, fired, defended by his fellow hack Sean Hennessey who in turn is fired by the stammering, sallow, cowardly Henry Mall, the paper's senior editor. By the end of the novel, we learn that Sean is a wealthy major psi guru and the soundless speaker advising Danny-the-finder to beware. Mall is an immensely powerful psychic foe who quite justifiably observes: "Mock if you like. An infirmity of speech isn't the same as an infirmity of mind" (p. 235). In the avalanching narrative, the absurdity of all three of these extremely rare supernormals just happening to work for years (eighteen, in Mall's case) on a minor trade periodical is left floating. Have the other two precognized the future in which Danny, in a sort of "Secret Life of Walter Mitty" come true, graduates to psi master class, and hence worm themselves in advance into his future office?

Perhaps the key to the mystery is the IQ and personality testing Danny underwent before being shoehorned into this employment for which he had no obvious aptitude or interest. If so, the conflicting interests comprise a conspiracy war of serious proportions. It might be simpler to assume that Blish lost control of an early work written fast and patched up with excuses and evasions.

$$\Psi$$

None of this really undermines the novel's standing as a major contribution to the emerging sub-genre of psience fiction. If some of the major players are wildly inconsistent, and most of the rest are slightly enriched pulp conventions (fraudulent medium Mme. Zaza and her daughter the snide but electrifying Marla, the Frater Hegemon and his cultish robed crew, the moronic criminal thugs, crankish Cartier Taylor of the Forteans, notable astronomer Sir Lewis Carter of the Psychic Research Society, the jolly non-psychic parapsychologist Dr. Todd), still the ideas are not just enunciated but embodied in dramatic action and remain fresh even after some seventy years. How does psi work? Can it be amplified? (Yes, by a device called a resonator, devised by the remarkable Sean.) What kind of universe do we inhabit, allowing psi to exist for some, perhaps potentially all?

Blish offers answers, not mere hand-waving, to all these central psience fiction questions. This is a true novel of ideas, however zany and in some cases scientifically erroneous those ideas proved to be—the most embarrassing being Dr. Todd's assertion that the new psi function "may be located in the four-fifths of your brain for which no function is known right now." He adds, perhaps more plausibly: "You're opening up new synapses, new impulse-channels from nerve-cell to cell. Many of them have never been used before, and, for all we know, they may still be in a highly primitive state" (p. 68). There are also mentions of positrons flying about, when he means anions—there'd be a lot of matter-antimatter annihilation explosions otherwise).

Consider some of the crucial moments where Danny and others unapologetically puzzle out some answers on the basis of advanced science. He has levitated a heavy table of Mme. Zaza's, with pain and a whirling sensation, watched it move across the ceiling before crashing down. Todd explains, drawing on one of Blish's favorite mathematical devices, the Blackett-Dirac formula:

> "That's the Blackett effect," he said. "You cut down the gravitational field of every atom in that table. Sheer centrifugal force from the Earth's rotation put

the table aloft, and Coriolis forces from the same source made it crawl along the ceiling...."
"Atoms? Then the whirling things—"
"Are electrons, of course.... All we need to do is implant full conscious understanding of the physical process in your cortex, and the last connection will be opened.... Here, look at this."
The paper said: $G=(2CP/BU)^2$. Danny looked at it.
"I don't feel any different," he confessed.
"Do you know what magnetic moment is?"
..."Let's see—it's, uh, the product of the strength of a magnetic charge and the distance between the poles?"
"Right. Now look back at the equation. Magnetic moment is what P stands for. U stands for angular momentum, G is the universal gravitational constant, and C is the invariant velocity of light. What I'm out to show you is that a magnetic field is a product of rotation on an axis, and that gravity is a function of it. Now if you'll remember that every electron is a tiny electromagnet, and figure in B as an uncertainty correction amounting to about zero point twenty-five—" [pp. 107–09].

Another equation, the original Heisenberg "indeterminacy" formulation, solves the rest of the problem:

pq–qp=h/2πi(1)
The q was the generalized coordinate ... p was momentum, and h was the Planck constant.... Momentum, at least, he understood; in terms of the psi powers it would probably be equal to the velocity of propagation of nerve impulses, which was, as nearly as Danny could remember, in the neighborhood of 60 feet per second. Change that velocity, and you change the amplification of the machine that ran on it, the brain, and hence the brain's range of detection.... Planck's constant was ... an impossibly tiny sum—6.547×10^{-27} erg-seconds ... it would be indetectable except in the realm of the electron. It was a quantum-value—
Which would vary in an infinitely overlapping series.
There it was, staring him in the face. Necessarily, h would change in value at a constant rate from frame to frame; it was, in fact, the difference which kept the frames separated from each other.
The timing principle! [pp. 160–61].

These "frames" are "sequences of the serial universe whose totality made up the main line" of spacetime history. This proposition from 1949 is borrowed in part from J. W. Dunne's *An Experiment with Time* (1927) but seems uncannily akin to the superposed alternative values of each uncollapsed particle's state vector, in Hugh Everett's Many Worlds hypothesis, not published until 1957. The frames are crucial to the plot, when Danny uses an egg-shaped resonator to tweak the settings of Planck's constant (or rather, variable) from one "sigma" setting to the next, flinging him into a sequence of universes, each further into a different and distinct possible future than the one prior.

Blish brilliantly embodies this mathematical-physics notion as a series of steps rising in a multifoliate staircase, endless alternative sets of steps bending away at right angles from every other: "Every time Danny went from one step to another, he would be snapped into another world. Not an unreal world, but not a real one either; simply a nearly-fictional world which could become fact only by being lumped with its sibling half-truths" (p. 207).

Ψ

Regrettably, here in what seems to most of us to be the real world, nobody has yet devised a resonator device that can spread apart the multiple worlds to allow premonitions and other psi activities. It seems unlikely that any of Blish's conjectures in this dazzling display of informed imagination will prove to be correct, but the very presence of psience fictional texts like this one can prompt fresh analyses and experiments by researchers into the paranormal. And provide a vivid sense of what such a transformed, psi-aware world might be like.

6.
James H. Schmitz, *The Witches of Karres* (1949/1966)

At the end of the forties, heading toward the golden days of psience fiction in the following decade fueled by the endless propaganda from editor Campbell at *Astounding*, the sub-genre speciated from its origins in fantasy to colonize the available modes and colorations. Not all such psience fiction any longer portrayed grim mutant heroes and heroines (slans, say) embattled against vast political and social forces, or sardonic mutants taking themselves out of the game (like Odd John), or even boyish super warriors with impeccable ethics (such as the Gray Lensman). In 1949, James H. Schmitz sold Campbell one of the fondly recalled "greatest science fiction novellas of all time," Hugo winning "The Witches of Karres." It was collected a quarter century later in one of the volumes of Doubleday's *Science Fiction Hall of Fame*, stories chosen by the members of the SFWA (Science Fiction [and latterly Fantasy] Writers of America).

That longish jape, considerably extended in 1966 into a novel with the same title (and sequeled in 2004—*The Wizard of Karres*, by Mercedes Lackey, Eric Flint and David Freer—and further in 2010: *The Sorceress of Karres*, without Lackey), was a curiously sweet, amusing adventure romp. Young Captain Pausert from the planet Nikkeldepain (he might be in his mid-twenties, although he often seems older) visits a world of the wealthy Galactic Imperium. In the empire slaves are readily bought and sold, and Pausert finds himself enmeshed with three stolen prankish child witches. These are Maleen, who seems to be about fourteen, her equally blonde little five- or six year-old sister, the Leewit, and their dark-haired suitably gothic middle sibling, Goth, a disconcerting nine-year-old. It's a charming if potentially creepy crypto romance, where we discern that one of these imps will surely, sooner or later, pair off with the captain, wrecking his plans to wed the beautiful and wealthy Illyla Onswud back on his home world. Finding these children suffering a hard time from their purchasers, the captain is obliged by his own ethics to buy them himself and take them away to safety, even though doing so is a major crime among Nikkeldepainians.

Their home world is Karres (does that rhyme with "heiress" or "caress"?) which proves to be an entire runaway planet of witches who use psi powers individually or *en masse* to create, scale-up and control the faster-than-fast Sheewash Drive. We rapidly learn by simple observation that Maleen has precognition and telepathy, the Leewit can smash precious vases with a piercing sound, and Goth manages to steal a swag of jewels from her former owner, presumably by teleporting them into her bag. Little is explained in data-dumps; what we and the captain learn is what we see happening, which is appropriate for a tale about childhood and its exit into early maturity. Goth, admittedly, does eventually announce that she has ear-marked the captain, but that nothing will happen for a couple of years, Karres time. "Be a bit before *I'm* marriageable age.... So I got it all fixed ... as soon as they started saying they ought to pick out a wife for you on Karres. I said it was me, right away; and everyone else said finally that was all right then—even Maleen, because she had this boyfriend" (p. 60).

The narrative begins as a picaresque in a style akin to Eric Frank Russell's, and continues in that vein. The old rustbucket former slaver ship *Venture 7333* still holds some power up her sleeve. Pursued by forces of evil and righteousness alike, Pausert can generally outrun or outwit his foes, avoid threatened years in prison and fearsome weapons, and if that doesn't work Goth will step in and activate the Sheewash Drive, at significant if temporary cost to her well being. At length, the ship arrives at Karres and the captain is introduced to Toll, the girls' mother, and learns that their father is his great-uncle Threbus from the planet with the ludicrous name—making the girls his first cousins once removed.

Returning home, he waits on orbit while the police of the Republic gain access, inspect his cargo (much of it inadvertently illegal) and he is chagrined to find that not only has he been spurned by his insipid girlfriend, but in the meantime she has married his rival. Luckily, and surely implausibly, Goth has hidden herself as a stowaway and manages an escape from further hazard using her psi powers. Off they go across dark light years with the Sheewash Drive flickering orange flame above a cone of bent wires, until Goth faints. The novel continues in comparable episodes of challenge and response, grace under pressure, shrieking or doughty girls, silence, exile, and cunning.

Ψ

In the belated but seamless extension of the novella, the captain reads a PROHIBITED notice in his voluminous stack of *General Instructions and Space Regulations of the Republic of Nikkeldepain* that somehow he's never bothered to study previously. The witches of Karres, it appears, had either "developed an alarmingly high level of secret technology or ... there was something downright supernatural about them.... There was grave danger of spiritual contamination." More, the women of the mobile world were "gifted with an evil allure.... Hiding under the cloak of so-called klatha magic—" (p. 63).

Here, at last, is a key to this psience fictional mélange. "Klatha," the captain recalls, "was a metaphysical concept—a cosmic energy, something not quite of this universe. Some people supposedly could tune in on it, use it for various purposes" (p. 63).

In short, and a decade before the first *Star Wars*, klatha was ... *The Force*.

Ψ

Pausert, like Blish's Danny Caiden, soon learns to access this force: "There was a feeling as if the universe had stopped for an instant; then a shock of alarm. His scalp began prickling as if an icy, soundless wind had come astir above his head.... Some klatha machinery was already in motion now and couldn't be stopped" (p. 71).

And he performs a minor miracle of teleportation, while insisting to Goth that he has no klatha ability. She demurs. "You got a lot of it! ... On Karres they all knew you had it.... You put it out so heavy the grownups were all messed up" (pp. 73-4). The captain learns that

> One didn't *produce* klatha. If one had the talent—inborn to a considerable extent—one attracted it to oneself. Being around others who used it stimulated the attraction. His own tendencies in that direction hadn't developed much before he got to Karres. There he'd turned promptly into an unwitting focal point of the klatha energies being manipulated around him.... [O]nly the younger children, using klatha in a very direct and basic, almost instinctive manner, weren't bothered by it. Adolescents at around Maleen's age level had been affected to some extent, though not nearly as much as their parents [p. 75].

It turns out, frustratingly, that the captain's miracle was actually performed by what we might call a poltergeist—a barely visible *vatch*, or playful klatha creature drawn in from some alternative universe, attracted by a disturbance in the local force.

A far more alarming *vatch* takes an interest in the captain and Goth in the midst of dangerous hugger-mugger which comprises the rest of the narrative. Convinced that the "small human" (Pausert, not the child) is an imaginary piece in a Game, like all other beings not *vatches*, it flings them and their starflight companions and foes from one threatening locus to another. It bears an uncanny resemblance to *Star Trek: Next Generation*'s mischievous semi-god Q, two decades later—so much so that the similarity hardly seems likely to be a coincidence. In any event, the caper proceeds with some ingenuity from one hazard to the next. Pausert and Goth best various sly plotters including the Agander, chief of the galaxy's dread buccaneers, match wits with interdimensional reptiles and the Giant-*vatch* overseeing this sport, are flung backward in time, learn that the captain's particular psychic gift makes him a specialist vatch-handler and master of klatha locks. All of this is great fun but reveals no additional insight into the purported psi force other than its handy and flexible plotting availability to the author.

Such is often the case with paranormal motifs in psience fiction of the strictly adventure yarn variety. Schmitz's effort here is more smoothly accomplished than many others, but it does not lead toward development of klatha/psi as a possibility more likely than fantasy's magic. In a 1955 stand-alone novel, *The Ties of Earth*, to which we shall return, Schmitz makes a more interesting effort to portray psi with discipline and (in a suitably melodramatic fashion, of course, since it was aimed at *Galaxy*) even realism.

7.
Alfred Bester, *The Demolished Man* (1952)

At the beginning and near the end of the 1950s, the brilliantly coruscating Alfred Bester produced two of the finest psience fiction novels ever published: *The Demolished Man* (first serialized in the ambitious then-new magazine *Galaxy*, starting in January 1952) and the even-more astonishing *The Stars My Destination* (also *Galaxy*, 1956–57). We shall return to the latter. The former is a vivid blend of crime novel and science fiction, set in a future where telepaths are not hunted and reviled, as in earlier psience fiction, but envied and rewarded for their gifts. This postulate creates a kind of narrative paradox, which is Bester's *raison d'être* for the story: how might crime, especially murder, even occur in a world of ESP "peepers"? If it is committed not in an unpredictable burst of rage but after diligent and clever scheming, can a killer find a way to evade a police force of trained and organized telepaths?

The Demolished Man was such an energizing shock to the sf readership of its day, chosen as the winner of the first Hugo award and a landmark ever since, that it might seem unnecessary to rehearse the plot and memorable characters. Still, it appeared nearly a lifetime ago, and the trope of police telepathy has been recycled by others ranging from Philip K. Dick's 1956 "Minority Report" to J. Michael Straczynski's *Babylon 5*—where a war criminal Psi Corps operator was given the name "Alfred Bester,"[1] a notable mark of disrespect presented as admiring *hommage*.

It might seem equally disrespectful to unpack the structure of Bester's artfully deceptive plot, but for present purposes it is the clearest path to the model of psi activating this novel. Some of these background elements were sketched more explicitly in the serial's opening than in the somewhat modified and pared-down book publication. Stylistically, this was a sound choice, now opening with "Explosion! Concussion! The vault door bursts open!"—which is a comic strip summary not just of billionaire Ben Reich's terrifying dream of the Man With No Face, who will pursue him in nightmare, but also a description of the book itself.

In 2301, Reich's heritage (or so he imagines) is as heir to Monarch Utilities' immense but failing interplanetary corporate empire, and this tangle of history is displayed in the abandoned opening chapter. In fact, as he unconsciously recognizes (and this tale is nothing if not Freudian to the bone), he is the bastard son of Craye D'Courtney, his elderly corporate rival. Using Lloyds' Executive Code, Reich offers merger with D'Courtney's cartel. The old man accepts in the same code, wishing to assuage his guilt at rejecting his son all these years. Despite his skills as a decoder of signifiers (we see him effortlessly position himself in time and space with a panoramic clock showing the equivalent hour on Venus, Earth, Mars, the Moon, Ganymede, Callisto, Titan and Triton), Reich misinterprets the reply and ragingly determines to kill D'Courtney before his own cartel is bankrupted.

Ψ

The *Galaxy* magazine version begins with a series of brief, mockingly mock-journalistic retrospective entries: fat, sallow, boring Edward Turnbul who in 2103 discovers the secret of antigravity and has it stolen from him. In 2110 Galen Gart finds that he and his recently deceased wife have been communicating telepathically, and when he marries bubbly 20-year-old Duffy they have a son whose "extractive recessive" psi genes gift him, too, with ESP, powering a ruthless capacity to blackmail and steal until he is murdered.

Several items are misplaced and forgotten but will play a part in Ben Reich's plans. Generations of Reichs and Courtneys clash, growing ever richer. Young Lorry Gart sets up as an uncertificated advisor for those with psychological problems, using his ESP abilities in an ethical way. A court case is decided in favor of an accused who objects to the use of psi to testify as a hostile witness against himself.

The *Geoffrey Reich*, ostensibly first crewed ship to reach the Moon, finds a crashed vehicle in the middle of a seventy-mile bed of valuable Haines' Stellite. An earlier space adventurer had crashed on Mars, chalking on the deck of his vessel *Die Kunst ist lang, das Leben kurz, die Gelegenheit flüchtig*, or Art is long, Life short, Opportunity fugitive. The impact, it turns out, dug a crater of "radiant magma" worth $40 million, and a D'Courtney tanker found the ship. (Both sources of mega-wealth are thus accidental and in effect unearned.) Ben Reich, when he learns of D'Courtney's windfall, is typically sardonic: "Magma Cum Laude." But this find is one of the sources of his imminent ruination. He can achieve his revenge only by slaying his rival in a way that the ESP law agents cannot unmask—achieve the first successful murder in more than 70 years.

In essence, then, this psience fiction novel is centered on the conflict between the psi-gifted and the felonious but brilliant and determined (apparently) non-gifted. And the open detective fiction element of the narrative is, somewhat unconventionally, revelatory to the reader in many important ways: the murderer is known from the start, and watched closely at every point, even when his Freudian unconscious meddles with what it can't allow him to know but we can see or guess if we are quick.

Reich's opponent is not his victim but the Police Prefect Lincoln Powell, a whimsical but forceful Esper 1st Class. After the murder, which D'Courtney's daughter Barbara sees before she flees traumatized from Reich, these two meet and both are impressed. Despite the rather heavy signatures of the names, Reich is not a hysterical little man with a bad haircut and ridiculous mustache. And despite his lack of conscience and the hatred and dread of his staff, he has wealth, power, good looks, sexual appeal and energy. Powell quickly cuts through the clever psi-blocking jingle Reich has sunk into his own mind ("Tenser said the Tensor"), but the psi cop knows that it will be nearly impossible to prove motive, method and opportunity—a conclusion all but guaranteed when investigators find that the dead man was on the verge of signing a merger with Reich. Where, then, is the motive for murder?

Already, we have a nicely woven blend of crime and psience. The plot is complicated by a series of failed attempts to kill Reich, by his murderous hunt for Barbara, and his terror at the repeated and exhausting dreams of the Man With No Face—who cannot still, after all, be D'Courtney. A key to Reich's inner conflicts can be seen: his own hidden, undeveloped or even blocked psi abilities. In an early psychiatric session with an inferior peeper (the only grade he dares to hire), a mental free association response to "Robbery" is:

> "Jewels—watches—diamonds—stocks—bonds—sovereigns—counterfeiting—cash—bullion—dort..."
> "What was that last again?"
> "Slip of the mind. Meant to think bort ... uncut gem stones."
> "It was not a slip. It was a significant correction; or rather, alteration" [p. 8].

Reich's masked thought, as we realize only much later, was "daughter"—Barbara, his sister. There are hints of incestuous longings, even though he has no conscious knowledge of the young woman or her relationship to him. At the moment prior to the murder, "a half-dressed girl bursts into the room ... nude under a frost silk gown hastily thrown on, yellow hair flying, dark eyes wide in alarm. He caught her while she fought and screamed.... Reich shook with galvanic spasms" (p. 49).

$$\Psi$$

His mirror image, Powell, faces even more blatant Oedipal (or rather, Jungian Electra) impulses when he takes Barbara, regressed by trauma to infancy but swiftly recovering her sense of self through recapitulated girlhood and adolescence, into his bachelor residence. Two-thirds of the way into the novel, Powell returns home and finds her with "a black crayon in her right hand and a red crayon in her left. She was energetically scribbling on the walls, her tongue between her teeth ... "Drawin pitchith," she lisped. "Nicth pitchith for Dada." Powell probes "through the vacant conscious levels of her mind to the turbulent preconscious" and beyond, to "the faint spicule of a star that burned with the hot roar of a nova.... Abruptly he was face to face with himself.... His image ... was nude, powerful; its outlines haloed with an aura of love and desire" (p. 118). "He followed the Janus image down to a blazing channel of doubles, pairs, linkages and duplicities to—Reich? Imposs—Yes, Reich and the caricature of Barbara, linked side by side like Siamese twins, brother and sister from the waist upward.... Barbara and Ben. Half joined in blood" (p. 118).

Much later, as she recovers, she enters an adolescent phase, with attendant hazards.

> "Please, Barbara, you're terrifying me.... I knew you so intimately as ... well, as a child. Now..."
> "Now, I'm grown up again."
> The urchin appeared again in her face.... He dropped the parcel and caught her in his arms.
> "Mr. Powell, Mr. Powell, Mr. Powell..." she murmured. "Hello, Mr. Powell" [pp. 187–8].

Her infatuation, and his, could have no serious outcome if she were non-telepathic since Powell, as a highly ranked member of the Esper Guild, is obliged to marry another Esper. Luckily, she does share the embryonic gift with her half-brother Ben, and it woke her from sleep when her all-but-mute father mentally called for help as Reich threatened him with death. The psi ability blossoms under Powell's loving attentions.

ESP utterly permeates the novel, whether recognized (the Espers are everywhere, inviting people at large to announce themselves and join this new elite) or hidden. Bester's literary skill is in making this capacity visible on the page, however implausibly. High level peepers adopt something like leet spelling of their names, which we take to be emblems visible only to the mind of a telepath: ¼Main, @Kins, Wyg&. Reich's tame Class 1 accomplice in murder, the racist psychiatrist Tate who wants the world ruled by Espers, was shown in the serial as T8. Dinner parties are an explosion of surrealist interpenetrating words/images/emotions/sensations that cross and interpenetrate in three-dimensional space rendered as two on the page. No example of such sportive psi has ever been reported by the psychically talented, but as a graphic metaphor it works supremely successfully (pp. 24–5, 27).

The moral plague that is Ben Reich, meanwhile, ruins or kills as it pleases. A poignant recurrent image is the lonely, shunned pawnshop owner Jerry Church, formerly Class 2 Esper, lurking outside gatherings of peers like one starved. He has been expelled for participating in another unprosecuted crime instigated by Reich. In the final stunningly orchestrated scene (later often appropriated by TV and movies), Reich is convicted and subjected to Demolition, the total scouring of his personality so that his powerful potential can be rebuilt in a more socially acceptable frame. This horrifying and literally mind-murdering punishment/rehabilitation casts a serious shadow over the ethics of Powell and his fellow Espers. We share with Reich the occlusion of his world, his universe, his very self. Undergoing a Mass Cathexis, akin to the far simpler psychokinesis deaths in "The Public Hating," he sees that the stars are gone, and nobody has ever heard of them. Then the planets vanish, have never been. No D'Courtney Cartel. No files at Monarch. The Moon has gone. The Sun has gone. There is no Paris. He sees The Man With No Face. Himself. D'Courtney. Both. It was ended, in Demolition (167–181).

Is this a plausible future for psience fiction? No, it is high-energy melodrama, playing on the dreads and solipsistic anxieties of the mid-twentieth century, hardly any reference to real parapsychology beyond the Rhinean term "ESP." The advanced brainwashing Bester allegorizes is not unlike the disgraceful MK-Ultra experiments of CIA in the U.S. during the Cold War and the Soviets in their own mind-smashing technologies. Even so, *The Demolished Man* is a remarkable and powerful feat of imagination, and a necessary allegorical strut supporting the arena of psience fiction.

8.
Zenna Henderson, *The People* (1952)

An Arizona elementary school teacher raised as a Mormon, Zenna Henderson (1917–83) is most noted for her extended story sequence about the psychic humanoid People crashed upon Earth. Created between 1952 and 1980, and published first in two volumes and then collected—plus the long-unpublished "Michal Without"[1]—in *Ingathering* (1995), the series comprises a remarkable saga of the ethical communal uses of psi. Critic John Clute notes in the *Science Fiction Encyclopedia* that

> decorous warmth infuses [and] inescapably marks *The People* tales, and her portrayal of women in unchallenged positions of authority was noted early and favourably by feminist critics. It is true that her patent decency sometimes overly reduces tensions and contrasts, but though this wholesomeness can be vitiating, the humaneness almost always shines through.

Because her inaugural story in the series, "Ararat," was published in *The Magazine of Fantasy and Science Fiction* in 1952 and established the back-story and tone of the three-decade venture, the entire arc of the sequence can be suitable addressed in this chapter. The title is of course a poignant reference to the Biblical mountain-top resting-place of Noah's Ark after the Hebrew god, Yahweh, drowned humanity's wretched first habitation, leaving only a small familial sample of people and animals to begin again on the depleted planet.

That biblical underpinning is reinforced by subsequent story titles and themata: "Gilead," "Pottage," "Wilderness," "Jordon," and so on. How these crashed aliens from another world happen to be entirely human except for their abundant psionic gifts is unexplained, except for the sense that in a Mormon universe life flows from the creative and destructive hands of "Heavenly Father," a transcendent human from an earlier cosmos, so of course they resemble us, as presumably do all other sapient beings in the cosmos. A Latter Day Saints source notes: "Joseph Smith saw for himself that the Father and the Son were two separate and distinct beings, each possessing a body in whose image and likeness mortals are created.... Gods and humans represent a single divine lineage, the same species of being, although they and he are at different stages of progress."[2] This is not to claim that The People series is based on LDS doctrines, but certainly there are illuminating similarities in that theodicy.

An astute online commentator, system administrator Russ Allbery, assesses with some precision the personal appeal of the People corpus to many smart, outsiderly sf readers:

The stories are almost uniformly about being lost, alone, misunderstood, hunted, and endangered, about thinking oneself insane or broken, and about the glory of finding another or a whole community who understands and welcomes you…. Henderson is covering well-trod ground in SF, but she does it with such a warm heart and such obvious joy and love for her characters that her telling seems fresh…. [*Pilgrimage*] is, to some extent, a wish-fulfillment fantasy: loneliness, isolation, and a sense of being different [from] all the people around may be cured by discovering one day that one is a part of a wonderful community of fellow aliens who can teach special powers.[3]

It is Henderson's deployment of these "special powers," or Persuasions, mostly versions of sf's canonical psi tropes (but not really parapsychology's), that makes her work especially salient to the megatext of psience fiction.

Like Krypton, Superman's birth planet, Home perished in a cataclysm but not before some star travelers from that world managed to escape. First they went to New Hope, then to Earth, where their vessel suffered a calamity of its own and many of the passengers and crew perished. Not all; we first meet the Group, who live in near isolation in the rocky and rather forbidding terrain of Cougar Canyon. Much of this back-story emerges only in later stories from the 1960s, perhaps because Zenna Henderson didn't know it at first. Her development of the alien settings is always somewhat vague and undernourished, because what she explores is not alien geography but the hearts of the lonely and dispossessed.

Even the collective and loving People of the Group suffer these pangs, recalling their lost, beloved home world. To those wounded souls who stumble upon them (like Lea, an Outsider whose adoption frames most of the first volume of testimonial stories), or are lost People sought out and embraced, misery is transcended into psychic fellowship, within a kind of indeterminate deistic faith, and determination to make a successful life in this new place where their psychic gifts blend miraculous opportunity with the endless risk of witch-hunts and cruel rejection by Earth humans.

From the outset, in the original and now post-frame tale, routine use of a specialized vernacular reveals that the psience is embedded deeply in the parallel culture. People who levitate (as they do whenever it's convenient, or for fun, once the power emerges at the age of five or so) are "lifting." More complex control of mass invokes "platting the twishers." Almost everyone can do sunshine by grabbing a beam of sunlight. "Of course only the Old Ones do the sun-and-rain … and only the very Oldest of them would dare the moonlight-and-dark, which can move mountains" (p. 32).

Some experts, like Karen, with the gift of "sorting," can unruffle and ease the pained confusion of the troubled (p. 33–4), a kind of do-it-at-home psychotherapy. The garden blooms with a plant from Home: *koomatka*, which survived the Crossing; the loss of others such as *failova* and *flahmen* indicated the coming Calling to the Power (or destruction) of the planet Home, where People had names such as 'Chell, Lytha, Thann, and rather oddly Simon and David. The visionary or precognitive, a rare breed, See such impending serious fate.

When we Earth humans were first encountered by the castaways from the flaming starship we were despised by the People as *undene*. Later, our ambiguous worth is recognized, and older adolescents may go outside the Group for university training, but the People tend to maintain a kind of rural ghetto existence. As in many psience fiction novels and stories, this is something between self-selected genetic apartheid and retreat from possible harm at the hands of the *undene*, as well as provision of a free and nurturing micro-environment.

Above all, the Group and others like it await another starship to rescue and carry them away to their own people. The irony of the second volume is that many prefer to stay with the replacement culture they have developed on Earth, even though they have no Healers. It is a recurrent tension at the heart of psience fiction, even the large portion where the gifted are radiation-modified mutant humans or evolutionary "hopeful monsters" such as the emergent group mind in Theodore Sturgeon's *More Than Human*.

Moving as many of these linked stories are, despite the dusty patina of their origins in magazine fiction of the 1950s and 1960s, the saga of the People remains essentially a religious fairytale rather than genuine interplanetary science fiction. Is it psience fiction, though? Certainly. Psi is what powers these children of the Power, even if it takes forms that are extravagant in terms of what we actually know about ESP, telekinesis/psychokinesis, precognition, and so forth. It remains an audacious and touching portrait of an appealing world, even a utopia of sorts, that might emerge from the chrysalis of the psychic realm—assuming that the fleeting and unreliable forms of psi that we witness are somehow upregulated and brought into conformity, in the future, with everything else science comes to understand about the cosmos.

9.
J. T. McIntosh,
The ESP Worlds (1952)

So far, most of the psience fiction works we have explored (other than the first two, neither of them noticeably generic) were written and first published by Americans. For reasons of the war and post-war economies, and differential populations, this is hardly surprising. It was more than two decades after Olaf Stapledon's *Last and First Men* appeared in the UK that a British magazine published a full-length psience fiction serial. Even then, with J. T. McIntosh's *New Worlds* serial *The ESP Worlds* (Numbers 16–18, July–November 1952), readers were confronted (without being informed of this fact) with a serial made of two stories jammed together by editor John Carnell without McIntosh's knowledge or blessing.[1] Even so, the narrative does follow several of the same characters on two quite different worlds where psi phenomena are common among the human residents, unlike back on Earth.

The result is not impressive. These were early issues of what would become the most consequential British sf magazine, but editor Carnell had not yet had much time to build up a reliable stable of writers, let alone a few brilliant innovators like J.G. Ballard and Brian W. Aldiss. This journeyman piece by McIntosh—who eventually would sell quite often to U.S. magazine and paperback houses, before giving up in despair—shows no sign of careful world-building or characterization, and his approaches to the phenomena of parapsychologists are frankly laughable. Running to some 38,000 words, it comprises a sort of primer in how not to write a psience fiction work.

Major Jeff Croner is Commanding Officer of the Universal Order Force stationed on the planet Noya, one of the two human-populated NO worlds; the other is Nome, also part of his command. He and his men have confronted the Noyan locals repeatedly, but been trounced each time by the strong young women of this matriarchal society ("for centuries their men have been drones," p. 5). These guerrillas used no bombs, guns or arrows in their resistance, simply teleporting into the arena of battle against Terran colonists and UOF troops, knocking their victims unconscious and vanishing away until the invaders withdrew with their wounded, bloody and bowed. The psi-mediated struggle had been going on for longer than Jeff could tolerate because the women of Noya naturally refused to deal with these lowly male opponents. The Major needed a woman to speak for the UOF, preferably a telepath, since ESP is another skill widely familiar to the natives. "We need a girl who is a woman and a half." HQ informs them "no female operatives of the necessary rank and experience available" (p. 5). But they are authorized to co-opt any suitable civilian.

By a stroke of luck, a ship has just landed with supplies, mail and a few new colonists—including seven women: four wives, two nurses, and a romance novelist, Janice Hiller. She proves to be blonde, tall and "if she hadn't written film scripts she could have acted them. She wasn't perfect; she looked too smart to be perfect." Thus, women in 1950s' popular fiction. This writer (or "girl") of popular fiction is not enthusiastic about her proposed role as intermediary, but after being frog-marched to Jeff's office agrees to risk her life for the cause as Acting Major Hiller of the UOF.

In a month of training, Janice sharpens her psi skills:

> She had had the usual telepathy training.... Fortunately she had great telepathic ability. Telepathy would never become general in the Earth system while there was so much variation in the abilities of different people. Everyone had a barrier, but not everyone could learn how to let it down to let thoughts in and out....
>
> But Janice, to pass as a Noyan, needed telepathic ability well beyond the average, and she had it. The Noyans had a spoken language, however, and it was very simple ... apparently used chiefly when for any reason telepathy was undesirable as a means of communication [pp. 8–9].

Despite their routine psi capacities, when Ala, one of the warrior women, "unexpectedly open[s] her mind," it is "the greatest shock of Janice's life" (p. 9). Even so, with instant aplomb the Acting Major responds to the requirement of reciprocity, "put[ting] in two seconds of the most concentrated thought of her life. She found in what she had learned from Ala the only possible answer," opening her own mind but passing herself off as a lowest caste slave. None of the Noyans noticed telepathically that actually she was a human from Earth.

A variety of alarums and conniptions follows, tribe against tribe, attack by a mental "stab" or "beam" from one of Ala's tribe that Janice blocks, discovery of a superstition concerning an imaginary "race in the south, the *ebru*, who were beings just this side of gods" (p. 13). Wily Janice immediately passes herself off as an *ebru*, taking advantage of a cliché from much 19th century adventure fiction.

While plenty of Earth and Noyan people have communication-class telepathy, clairvoyance seemingly does not exist, At length, Janice discovers the secret of the Noyans' teleportation: it doesn't exist either. Actually the women have marvelous hypnotic skills. Faced with the UOF foe, they simply render themselves invisible via hypnosis, approach each armed hi-tech enemy with a stout cudgel in hand, beat them unconscious, and run away, invisible once more. Apparently the future star-faring Earth has never developed radar, military grade cameras, satellite surveillance, drones, or even those small body-worn cameras that police in our own primitive time are obliged to don. Admittedly, the triumph of miniaturized technology was not obvious fifty years in advance, but still this is a depressing failure of imagination. With a moment's more forethought McIntosh's whole plot would have shriveled and blown away off the paper.

The final two segments of the *New Worlds* serial shift their locale to the other NO world, Nome, where Commanding Officer Croner is again attended by Major Hiller, revealed as the female UOF officer Jeff had requested but been denied. The romance novelist attribution was a cover story, but one supported by the fact that she actually was a romance novelist as well. Presumably her psi powers managed to keep the lid on this fact among other telepathically gifted humans on both Noya and Nome. In any event, the story limps along into further absurdities. Nome is a planet obsessed with Games, which, as in various of Jack Vance's later and vastly superior sf novels, carry not only potential prestige of Olympic proportions but also risk of instant death for failed contestants. This proves to be a strategy for thinning the population of this marginal world while selecting for psi acuity. The citizens would never tolerate the forty million players murdered each month (the Games are also rigged, of course), because they understood the number to be a mere five million. Perhaps this is McIntosh's ironic jab at unheeded road fatalities in our cultures, or the cost of free access to guns in the United States.

No fresh insights into the psi capacities of the NO worlds is forthcoming, nor how the seriously psi-gifted Janice will cope with her skills in life back on Earth. Jeff assesses her: "He gave her one hundred per cent for the way she looked in her subtly indecent Noman outfit, ninety per cent for the way she had handled things on Nome, and a hundred per cent for the coolness that had never cracked—two hundred and ninety per cent in all. It was too much for one girl" [Part 3, p. 94].

Janice, presumably picking this up telepathically, would have far too much coolness to go hypnotically invisible and beat the oaf about the head with a cudgel.

10.
Theodore Sturgeon, *More Than Human* (1953)

"The man has *style*" is how writer, critic and sf editor Judith Merril opened her encomium to Theodore Sturgeon in the tribute issue of the *Magazine of Fantasy & Science Fiction* for September 1962. Indeed, Sturgeon's fluent, sinuous, knowing style had marked him almost from the outset of his career as a major talent in science fiction. But what characterized his output even more markedly were his intense explorations of love and hatred, pain and joy, loneliness and the search for its alleviation in contact and empathy, and the utopian as well as potentially devastating effects of scientific research at the boundaries of the known, especially psi connectivity.

He was just 44 and, tragically, even then nearly at the end of his most productive career when that issue of the magazine appeared with his youthful, faunish (and allegorically horned) depiction on the cover, one of artist Ed Emshwiller's finest renderings, surrounded by the iconography of his most famous works. By 1985 he was dead, his best writing almost throttled for two decades by recurrent bouts of crippling writer's block. Despite these limitations, his most memorable work was steeped in the impulse and iconography of psience fiction, especially his astonishing novels *More Than Human* and what is now properly called *To Marry Medusa* (an expansion of a *Galaxy* story of the same title, for many years scarred by a publisher's ugly and inappropriate title, *The Cosmic Rape*).

Granted, his short stories and novellas were Sturgeon's true *métier*, collected and slightly annotated in thirteen handsome large volumes from North Atlantic Books (1995–2010). The long reaches of a mature novel were beyond him. The two novels cited above are not traditional in form; *More Than Human* is a composite expanded from Sturgeon's notable *Galaxy* novella "Baby Is Three" (1952), bracketed by a new introduction and coda, each of equivalent length and remarkably different yet complementary voice. "To Marry Medusa" appeared in magazine form in 1958, and in Dell at a brief 160 pages the same year, itself a kind of concoction of fragments displaying the wounded varieties of people before an alien invasion and almost immediate human transcendence into joy, mutual love, and a kind of triumphant global self-theophany. It is no anomaly that the long Sturgeon story featured in that 1962 *F&SF* tribute was titled "When You Care, When You Love."

Sturgeon's command of stylistics, colored almost always for good or ill by shades of his early indoctrination in the pulps, is not incidental to his psience fictional concerns. His diverse characters enact, with the verbal precision of his telling, their perceptions, deeds and utterances, their yearnings and achievements, the quest for connection beyond the body or the abstractions of reductive intellect—while retaining intelligence and a hunger for physical embrace. Many of these characters are frankly grotesques, especially in *More Than Human* with its quasi-mutant outsiders with strange deficits as well as compensatory psi powers.

Consider Baby, titular figure in the inaugural novella. Apparently an extreme trisomy 21 (Down syndrome) infant, but probably afflicted by, or gifted with, some unknown kind of chromosome variant, Baby is stunted and always will be, dull-eyed, speechless although later gestural, and yet telepathic in such an extreme degree that he is a living Google archive, with the search engine advantages as well. His intelligence is, so to speak, orthogonal to ours, and so is Lone's, his idiot protector who with Baby's help invents and then discards an antigravity device. All human babies can "hear" each other's thoughts, we are informed, and so can Lone as an impaired adult:

> Without words: *Warm when the wet comes for a little but not enough for long enough.* (Sadly): *Never dark again.* [etc.] … *It all rushes up, faster, faster, carrying me.* (Answer): *No, no. Nothing rushes. It's still; something pulls you down onto*

it, that is all. (Fury): *They don't hear us, stupid, stupid.... They don't, only crying, only noises.*

Lone, almost a wild animal, encounters two young women, Alicia and Evelyn Kew, and their demented and depraved fundamentalist father. The idiot finds innocent Evelyn and "from her came the call—floods of it, loneliness and expectancy and hunger, gladness and sympathy ... their silent radiations reached out to each other, mixed and mingled and meshed.... They sat close together" (pp. 15–16). It is not physically sexual, but a moment of revelation and transcendence for them both, the kind of epiphany that will recur in various mixed and blended voices through the novel, despite obstacles and confusions and denial.

Finally a greater-than-the-sum-of-the-parts comes together: Lone the telepath with eyes that seem on the verge of spinning, and encyclopedic Baby, and sportive, playful young black twins Beanie and Bonnie who can teleport, and runaway Janie Gerard the PK/telekinetic, and mad, resentful, selfish Gerry Thomson from a cruel orphanage, fated to replace Lone, and finally Hip Barrows, engineering genius ruined by his detestable medico father who just wants to be popular but never, really, makes it.

Together, they bond into a unity dubbed *Homo gestalt*, a small composite and powerful collective mind that remains, for all its satisfactions, alone. What is the nature of their new entity? What responsibilities and goals do its parts share? This is the moral—or rather, in Sturgeon's analysis, the ethical—issue that is resolved when righteous Hip become the conscience of the new species.

Here, then, is a novel entirely orchestrated around the theme of paranormal abilities, some never reported in reality, others a sort of hypostatization of known psi effects: the levitation, mind reading, the mind-over-matter (which is wonderfully and pungently exemplified when young Gerry is brought into the incomplete gestalt's hiding place and finds his bowels calling for urgent attention: Janie the PK agent solves the problem. "Suddenly I grunted and grabbed my guts. The feeling I had I can't begin to talk about. I acted as if it was a pain, but it wasn't.... Something went *splop* on the snow outside" (p. 92). This is not an effect discussed in any of Dr. Rhine's books.

One element of classical psychic research that is not mentioned in *More Than Human,* nor in most psience fiction, is postmortem persistence or postmortal reincarnation (a topic we shall defer to Appendix 2, below). In this case, perhaps that is because any instance of *Homo gestalt* will be, as Baby tells them, immortal. Even with his special gift, Baby will eventually die, and so will they all—but each will be replaced by another suitable human. Still, the loneliness that has driven them together remains. Perhaps it is inevitable, since this is after all a novel of aspiration rather than an experimental report, that at the moment of their coalescence with Hip as the necessary "*prissy one who can't forget the rules*" (p. 233), as Janie smilingly puts it, a barrier falls and Gerry-as-the-head of this small *gestalt* finds himself addressed by the many others who have come before: "a laughing thing with a human heart and a reverence for its human origins, smelling of sweat and new-turned earth rather than suffused with the pale odor of sanctity.... He felt a rising, choking sense of worship, and recognized it for what it has always been for mankind—self-respect" (p. 235).

Ψ

What might be waiting beyond *H. gestalt*? Why, a telepathic human Hive Mind. That is the psience fictional transformation *par excellence* (or at least for those who do not shudder in horror at the idea) for our species. And not just the psychic opening and bonding (the "bleshing," or blending and meshing, as it's named in *More Than Human*) of all our own defiantly individualistic kind, but of the entire galaxy, perhaps the universe. It is a notion known outside science fiction as the Noösphere, made famous at least briefly in the 1960s by Pierre Teilhard de Chardin, S.J., and familiar before that, if without that memorable term, to readers of Olaf Stapledon.

What precipitates this apotheosis is not meditation or spiritual mysteries but infection via a kind of swiftly multiplying spore introduced by invading aliens, the medusas of *To Marry Medusa* (1960). In our ocean biology, a medusa is a kind of bell-shaped or umbrella-like tentacular sexual jellyfish (technically, a cnidarians) that takes its turn in the life cycle with an asexual polyp. The intergalactic version (two and a portion galaxies, in fact) becomes linked to a loathsome, hatred-aching derelict named Gurlick who swallows a spore (Eucharist-like) in a burger.

The Medusa scans his wretched mind and takes steps, with Gurlick as its pain-and-sex driven creature, to begin the automated construction of a vast number of transformative machines. The world continues in ignorance of its impending fate, and Sturgeon thumbnails some of these people, creating something like a network of exemplary fictive characters that foreshadow the hive that will soon encyst them. Like the cast of *More Than Human* and many other Sturgeon works, they tend to be sad or wicked people, most of them warped or wrecked by the poisonous emotions of their childhood.

At a critical juncture the Medusa's planet-wide psionic machines activate and humankind becomes instantly a Hive mentality—each retaining individuality yet conscripted into the task of destroying the machines and readying for attack. When it comes, instead of absorption into the galactic entity, "Humankind became the creature; flooded it to its furthermost crannies, drenched its most remote cells with the Self of humankind.... The Medusa was alive as never before, with a new and different kind of life" (p. 154).

And here, finally, is the remade galactic ecumenism, with its rehearsal of the non-sectarian or even religious spiritual phrasing from the end of *More Than Human*: "So ended mankind, to be born again as hive-humanity; so ended the hive of earth to become star-man, the immeasurable, the limitless, the growing; maker of music, poetry beyond words, and full of wonder, full of worship" (p. 155).

Ψ

Sturgeon had been one of John Campbell's discoveries at the dawn of the Golden Age, until his perspective firmed into a humanism alien to *Astounding* but more at home in the *Magazine of Fantasy & Science Fiction* or *Galaxy*. In work from the same period to which we now turn, by Mark Clifton and his co-writers, more pedestrian but surely more appealing to the readers of Campbell's psience fiction, grand hopes for a merging of psi and science are also expressed. In the generations that have elapsed since that time, other visions of transcendence have arisen, notably expectations of more-than-human computer mentalities and perhaps a technological Singularity as impenetrable as any of Sturgeon's or Arthur Clarke's grand visions.

11.
Mark Clifton and Frank Riley, *They'd Rather Be Right* (1953–1956)

Labor relations psychologist Mark Clifton (1906–63), today almost forgotten, remains by some measures the archetype of a Campbellian psience fiction writer, initially in collaboration with archeologist Alex Apostolides (1923–2005) or columnist and travel editor Frank Rhylick, a.k.a. Riley (1915–96). Clifton claimed to have conducted at least 200,000 "man to man and off the record" worker interviews before retiring after a heart attack in the early 1950s.

The novel *They'd Rather Be Right* first appeared in three segments, all in *Astounding*: two short stories, "Crazy Joey" (August 1953) and "Hide! Hide! Witch!" (December 1953) by Clifton and Apostolides, and the four-part serial that followed directly on, *in medias res*, from that initial narrative, by Clifton and Riley (August-November 1954). In its first book appearance, the short stories were omitted (suggesting some refusal on the part of Apostolides to allow reprinting). Subsequent editions, sometimes under the title *The Forever Machine*, included one or both of the short prequels.

In any event, the second Hugo (the first having been won by Bester) was awarded for the magazine version. Because *Astounding* provided a sort of tally of reader votes month by month, we can see that the first part of four was beaten (if only by a small excess) by the instantly famous and controversial Tom Godwin story "The Cold Equations." Campbell went out of his way to comment after Part II that

> "They'd Rather be Right" seems to be doing very well for Clifton & Riley; the personal word-of-mouth reports I've gotten seem to be even stronger in its favor than the letter-votes—though I can't annotate them in the Lab here. I'll be interested to watch the results on the last two parts—when the backlash of "Bossy's" inability to help those who think they don't need help shows up! ["The Analytical Laboratory," *Astounding*, December 1954, p. 120].

Indeed, the last three parts were voted in first position.

Within a few years, its status fell precipitously, and the book was all but forgotten. In 2008, the *Guardian* newspaper commentator Sam Jordison excoriated the novel[1]:

> The challenge here is to provide an adequate impression of the pain of reading *They'd Rather Be Right* without drifting off into ludicrously extended similes about banging my head repeatedly against a wall of spikes or watching omnivorous ants munch me from the feet up. It's not just Dan Brown or Jeffrey Archer bad. This is a whole new arena of the appalling and you really have to have been there to understand.

He added: "*They'd Rather Be Right* is generally held to be the worst ever winner" of the Hugo award—although it is necessary to note that Barry N. Malzberg, curator of *The Science Fiction of Mark Clifton*, demurs with regret from this easy dismissal: "the most controversial writer of 1953 ... [he] was to end his life obscure not only to the American reading public but to the tiny, hothouse world of science fiction."[2]

Malzberg draws attention to a strain of bitterness and clear-eyed skepticism evident from the start. Clifton's first published story ("What Have I Done?" *Astounding*, May 1952) "stated at the outset of his career the belief that humanity was inalterably vile" ("Afterword," p. 294). One might readily gain the impression from his own considerable oeuvre that Malzberg shares this view. What is intriguing, though, is the missionary zeal that drove Clifton's writing. His first genuinely appealing story, "Star, Bright," appeared in *Galaxy* two months later. Although editor Horace Gold, famous for meddling with his author's texts, enraged Clifton by butchering the story in ways no longer discoverable, the tale was surprisingly sweet and hopeful.

The narrator has a "bitter confirmation" of his worst dread. His three-year-old daughter, Star, cuts and pastes a Möbius strip—a sealed band with a half twist in its curve—and with a crayon draws a line all the way along the strip, proving that it has only one surface, not a distinct outside and inside. Recalling his own cruel childhood, he sees this brilliance as an affliction: "Star was a High I.Q. ... A parent must teach her to compensate" by learning to pass as normal. It turns out that Star is already quite aware of this social necessity, and manages to pass as a dear little ordinary child even as she and an equally brilliant male friend teach themselves unusual tricks, such as time travel to the past (bringing back with them a brand new ancient coin), telepathy, and eventually zipping into parallel realities.

Star happily declares that she is a Bright, while her poor genius father is just a limited Tween. The rest of us are Stupids. Despite his defects on this preposterous scale, the narrator teaches himself to twist a four-dimensional tesseract mentally through higher closure and vanishes in pursuit of his wayward daughter and her Bright companion. This fairytale is, in its way, the ultimate version of the "Fans are Slans" self-mocking but self-gratifying mythos. Oddly, as its intensity suggests, this seems to have been analogous to Clifton's own picture of the world. Is it psience fiction? Not exactly, in this early story, but certainly the transcendent powers invoked are entirely those of the mind. We shall return to this ambiguous theodicy of the *Übermenchen*.

Ψ

Crazy Joey Carter is a sort of mutant quasi-Bright, tormented by his 1950s' McCarthyite head teacher and brutally stupid father—a trope that was still vivid in Uri Geller's *Ella*, in 1998—when he carelessly reveals his gifts. Taken to Steifel University (note the justified pun) where his father works as janitor, he is examined by young Dr. Martin who finds that Joey is telepathic, hides this fact from his boss, ultra-orthodox Dr. Ames, buckles and lets Joey down. He does salvage this cowardice with an anonymous letter revealing Joey's psi ability to Dr. Billings, dean of psychosomatics at Hoxworth University. Billings monitors Carter's progress, and to everyone's astonishment gets him a scholarship to Hoxworth. Now, pestered and threatened by another rigid McCarthyist, Billings seeks Joey Carter's help in designing a synthetic brain—what we would call an artificial intelligence but constructed of organic simulations of human neurons. Joey admits that he is also precognitive, but imperfectly: "I often see seconds or minutes ahead. Occasionally I see days or weeks but not accurately. The future isn't fixed. But I'm afraid of this thing" (*The Science Fiction of Mark Clifton*, p. 189).

The plot continues in a generally pulpish way, with Billings and his cybernetics colleague Dwayne Hoskins at last building the machine they name "Bossy" but finally crushed by the Board of Governors, and their AI dismantled: "they were powerless against the most ignorant of men. Against the most primitive flares of superstition and dread of the unknown, they had no defense. Weakly, in such a situation, they would try to explain, to reason, to appeal to rationality and logic—against minds preset against all explanations" (p. 212).

At the end of the second decade of the 21st century, this characteristic analysis in the school of John W. Campbell has proved to be sickeningly prescient.

The answer, apparently for Clifton as well as Billings, is that "man was just biding his time, slowly evolving, that psionics would mark the next stage, that [history's cycle] was a spiral and not a circle" (p. 214).

That much was prologue in these two Crazy Joey background stories to *They'd Rather Be Right*, which pursues the claim that people have been hamstrung into a stupid acceptance of their own subservience and that even industrialists and academics could not break through into creativity because they would defend to the death (usually of martyrs, but also entire cultures) their own indoctrinated biases and mental constraints.

This somewhat Ayn Randian doctrine is unfortunately illustrated in the novel by the reconstructed and super-intelligent Bossy's discovery that healing each cell in an aged and damaged body by psychosomatic means will induce rejuvenation—so long as the subject is prepared to abandon every arbitrary, superstitious, prejudiced and false belief. This has the disadvantage of inevitably losing almost all memory in the repaired and rewritten brain. Here is a form of Bester's Demolition, but seen from the point of view of victims of institutional and political stupidity and eagerly embraced. Of course, these renovated *Homo sapiens* also gain new mental powers: "esperance."

One of them, an aged and cynical former actress and for decades whore, Old Mabel, is returned to ravishing beauty and grace. Her enhanced mind, freed from "single value logic," is brutalized for a time by an onslaught of horrific mental chatter from a population of the unregenerate. "She was totally unable to adapt to a society which permitted the frustrated and psychotic to set up the laws and mores of behavior which resulted in the mass crippling of the whole human race" (Clifton and Riley, Oct 1954, p. 122). With Joey's aid she finally learns to defend herself. The story, meanwhile, follows all the paths one would expect: a canny billionaire, Kennedy, who steps in as a bulwark against the fake-news mass media and the sclerotic government, corrupt medical and similar authorities. The finale is a sort of heavily abbreviated version of a John Galt speech from Rand's *Atlas Shrugged*, but with multi-valued (and defiantly anti–Aristotelian) logic and something like Dianetics "clearing" as its premises. Joe speaks to the television cameras of the world:

> "Bossy is just a tool. Bossy can answer your questions, but only if you ask them.
> "There is another even wilder misconception. It has been said that Bossy is a soulless machine, and man, being guided by her, will become likewise no more than a soulless monster…
> "There is not now, there has never been any real issue between science and

faith. Both strive for the same identical goal; both seek comprehension; both wish to benefit man that he live happier, healthier, more harmoniously with himself and his neighbors" [Clifton and Riley, November 1954, p. 131].

Rand would certainly have had trouble with this acceptance of "faith," and probably Clifton would too—but he was a realist and knew how to avoid beating his head against a wall, even as he sought to guide his readers toward a cleaner, less cluttered search for truth. The appeal was obviously there, allowing this novel the second Hugo award, in 1955.[3] Its often heavy-handed literary crudeness might make us wince at that endorsement, but *They'd Rather Be Right* conveyed with some force a portrait of psi as a side-product and facilitator of superior thinking freed from antediluvian stereotypes and ancient sacred metaphors taken literally. For the psience fiction emerging from under Campbell's post–Dianetics wing, it was a kind of populist antipopulism finding expression in the new metaphors of pop parapsychology.

Shortly, as we shall see next, Clifton would discard Riley as co-author and begin a series of far more light-hearted but equally committed tales about the paranormal.

12.
Mark Clifton, "What Thin Partitions" to "Remembrance and Reflection" (1953–1958)

Between 1953 and 1958, Mark Clifton (solo except for the first tale) published four linked stories, and then later serialized a half-hearted invasion-from-space addition in 1961–62 in the less prestigious magazine *Amazing*, as *Pawn of the Black Fleet* (later, in paperback, the even drabber *When They Come from Space*. All centered on Ralph Kennedy, not by any means the same tycoon Kennedy from the Bossy stories, an industrial psychologist with much the same sort of job, skills and skeptical intelligence as Clifton's. Unlike the writer, presumably, this Kennedy's consultancy skill turns up a number of genuine psi adepts, including one who wrongly supposes he's a fraud.

Note that the phrases from which the story titles were drawn, quoted from Alexander Pope's *An Essay on Man: Epistle 1*, are curiously rearranged: the first is applied to the fourth tale, the second to story three, the third to the first story and the fourth to story two. Since the titles motivate the stories, always in an unexpected way, this turns out to make sense.

A couple more powerful pieces appeared elsewhere, and an unsuccessful novel, *Eight Keys to Eden*, in 1960. By the end of 1963, Clifton was dead.

Ψ

In the initial story, "What Thin Partitions" (September 1953), written with Alex Apostolides, Ralph Kennedy (no doctorate) is Personnel Director at Computer Research Inc., and on the same level of seniority at management conferences as other heads of department. This might seem a grim setting for a buoyant psi story, but the low-key guided tour through Kennedy's field of competent human expertise turns out to be at least as enthralling as any insightful description of a day working inside a nuclear plant or psychiatric institution for the terminally unstable. Part of this is due to the charming dynamic between Ralph and his admirable secretary Sara, always at the edge of sardonic disrespect but completely on top of her tasks.

No devotee of psionics, Kennedy deals with the irruption of a poltergeist child, isolated little Jenny Malasek with a Slavic mother "from the old country." Jenny's frustration trashes his office but the psychologist is calm and open to unusual explanations, especially when he finds a cylinder filled with a new chemical storage material sitting against one corner of his ceiling. It is designed to store impulses, and evidently psychokinetic antigravity effects can be stored handily. How to make the most of this serendipitous discovery? Why, bring in a Swami.

Ψ

The immediate sequel is "Sense from Thought Divide" (March 1955). Kennedy's common sense assures him that Swamis with levitating powers are frauds—even if they are provided by the Army's new Poltergeist Division. The child was a special case: "To a child who never knew anything else … one who had never learned to distinguish reality from unreality—as we would define it from our agreed framework—a special coordinate system might be built up" (p. 92) with which any metaphoric phrase might be instantiated literally. This is a recurrent theme in Clifton. Most general biases and prejudices might narrow and harm us, but some, in special cases, might unlock strange abilities.

With this conjecture in mind, Kennedy provides the Swami with some of the impulse storing capsules and invites the Swami to conduct a séance, which the fraud is happy to do. He become unhinged, of course, when the test objects start cavorting about the room, activated by his genuine parakinetic powers. Kennedy explains to the shaken man:

"I've been trying to build up a concept of the framework wherein psi seems to function.... It isn't something you can turn on and off at will. Aside from some believers and those individuals who do seem to attract psi forces, we don't know, yet, what to wrap around what.... You're to keep a supply of these cylinders near you at all times. If any psi effects happen, they'll record it" [pp. 111–12].

And serve with their antigravity and other oddities as the basis for both further research and for military or space programs.

Ψ

After a pause, "How Allied" (February 1957) was also published in *Astounding*. Five years after Sturgeon's "Baby Is Three," the seed of *More than Human*, we find Ralph dealing with the "*Gestalt* empathy" (p. 124) of a hive mind that names itself George. Comprising five healthy candidates for Computer Research Corporation, "fresh, young college grads ... all been turned out by the mass production education machine," they are seeking a company that "hires oddballs.... People with unusual talents" (p. 117). Ralph gives them a wry grimace but takes them on for "Something that five unconnected guys couldn't do, but George could do—" (p. 121).

It is a disaster. The five hires go to different parts of the plant, but because they share the tacit knowledge that each develops independently, the workflow improves to a level never imagined. Kennedy, chilled by this realization, wonders "Where did empathy leave off and telepathy begin?" (p. 125). He discards a little psi machine that he was playing with before he met the young men.

A military hearing is arranged to drum the gestalt out of the plant: a general and a "gorgeous Pentagon colonel," an admiral and an equally gorgeous Navy captain, a pair of Air Force brass, another colonel and his major. Almost at once, they begin to act like fools, competing at headstands, in the middle of the long table, jeering at each other like children, under the command of jesting George. As the five leave, "For one incredible instant George took control of me, and I shared the wondrous delight of being, belonging, the ecstasy of being something beyond human. Then he released me as the boys filed out the door, and [I] was left grubby, incomplete, ineffectual, bumbling—alone" (p. 143).

There's the theme again, found in so much of the transcendent imagery of psience fiction: the loneliness of a single, severed mind, forced to guess and bluff with others who are one's only extension of self in a world born from a meaningless vortex of cosmic gases. For science fiction readers of that time, half a century ago, mocked and derided before the arrival of "sci fi" on every screen, sharing the bliss of the sense of wonder was precisely an experience of "wondrous delight of being, belonging," if only for a moment before

12. Mark Clifton

reality returned. For a closeted gay man like Mark Clifton, it was probably even more agonizing (see immediately below).

Ψ

Curiously, the final story, "Remembrance and Reflection" (January 1958) appeared in the *Magazine of Fantasy and Science Fiction*, a more stylishly astute outlet than Campbell's magazine. *Astounding* took only a single further story from Clifton.

With the antigravity problem on its way to being solved by psychic means, Ralph is conscripted to a program to build a spaceship. He is offered help from the government.

> "The less help I get from the government, the better," I said drily. "Their ideas of merit are more concerned with a man's sex life, and what idle remark he may have made in an off moment twenty years ago, or whether he was ever arrested by some moronic cop. The more we get to be like Russia the less I like it. Have you ever noticed, Henry, that when a man takes on an enemy, he also takes on the characteristics of the enemy?" [p. 160].

Alan Turing had been persecuted to death less than four years before this story appeared in the *Magazine of Fantasy & Science Fiction*. Were his extraordinary war contributions, and his homosexuality, widely known by 1958? It is poignant to consider Clifton himself choking with rage in this global atmosphere of truly stupid malignancy. Could something akin to a psi global consciousness field help explain the rapidity with which this viciousness against gays has been changed in the last forty or so years? Probably not, but it is tempting to wonder if Clifton hoped for such a widespread surge of moral insight.

Ψ

In her introduction to his selected short work, Judith Merril mentioned that in their voluminous correspondence "he was writing with excruciating honesty out of his personal beliefs and ideals. Among these, of course, were his years of study of 'extrasensory' or 'paranormal' phenomena—and his years of *practice* of one particular ability—a sort of hyper-empathy which he called 'somming' (from *somatic*) and consisted of *experiencing* the other persona's somatic awareness" (pp. xiii–xiv).

He believed there was a slowly growing group of psi-sensitive people he called "emergees," those like Star Bright. He told Merril: "That is why I started writing science fiction; because the emergees read it. It is the only place they are able to find thoughts to match their own…. The emergee *may* not be one individual at all. The complete emerge *may* be a somgroup" (p. xvii, June 24 and September 9, 1952).

How curious close this is to both Zenna Henderson's People and many other groping intuitions (or are they just self-comforting wishes?) of psience fiction.

13.
Wilson Tucker,
Wild Talent (1954)

At a time when full details of the U.S. government's two decades of a highly classified research and operational psi program is finally available, Wilson Tucker's prescient novel of a telepath in government service is suddenly of particular interest. No mutant superman, Paul Breen realizes in early adolescence that he possesses the ability to detect the thoughts, moods and desires of others. Witnessing at the age of thirteen the murder of a G-man in a Chicago alley, and somehow intuiting the names of the two men who shot him dead, he ingenuously mails this information to the U.S. president.

In 1941, now a trained film projectionist, Breen is horrified by a Boris Karloff movie featuring telepathy and locates two books by J.B. Rhine, one by (fictional) Dr. William Roy, *Studies in Psychokinesis*, and a psience fiction novel, *The Time Masters* (by the real Wilson Tucker).[1] These teach him about not only telepathy but also precognition, which he has already experienced, telekinesis and teleportation.

Years later, as he is conscripted into the Army, his fingerprints from the letter are flagged and Breen is taken into involuntary servitude for his embattled nation. Unlike the real military remote viewers of the Star Gate program, after being identified as psychic Breen is eventually confined within a comfortable but massively guarded Maryland manor resembling Britain's Bletchley Park Manor House where Nazi codes were decrypted at about the same time. (Curiously, Bletchley's highly secret role was not publicly known until 1974, two decades after publication of *Wild Talent*.[2])

Tucker's naturalistic treatment of this unprecedented capture of a man said to be as Cro-Magnon was to Neanderthal is oddly compelling, slowly and convincingly building a frightening sense of persecution and entrapment. Hints develop that this history is not quite our own. In July 1945, Breen is taken into custody from his unit by an FBI and a CIC (Counterintelligence Corps) agent at the time of the Potsdam meeting that followed the victory of Allied Forces over the Axis, although this crucial VE triumph is not mentioned; the war continues, but only in Japan. While nobody in the FBI or CIC has ever heard of a newly developed weapon known as an *atomic bomb* (Part 1, p. 47), one is exploded over Hiroshima in August 1945 (Part 2, p. 103). The saga of Soviet vs. Western spies and nuclear prowess follows, but now the West has Breen as a secret weapon who can overhear the reports of U.S. spies and maximize their safety and effectiveness. He is chosen finally as one end of a spy relay proof against detection. Any failures are due to counteragents, and at least one of these turns out to be proof against Breen's interceptions.

For there are limitations to his ESP. It eventuates that Paul can read the minds only of those he has met or, preferably, interacted with. What of an informant who speaks another language, one he has never encountered? Makes no difference; he detects it, but "hears" or "reads" it as English. "'*Esprit fort*,' Paul quoted from [the CIC man's] mind. 'I am a free thinker. I am a strong-minded man.'" (Part 2, p. 123). His precognitive clairvoyant abilities signal danger but fail to prevent him from being shot (Part 2, p. 128). A female scream, detected mentally, suggests that there is another telepath in play, although he fails to trace her. From this point on, the story plods along to a rather tame ending in which Breen finds the woman who screamed in his mind as he was shot, falls in love with her, learns that she is not only psychic herself but has two brothers cut from the same preternatural cloth, and escapes with her from their captivity in Maryland after causing two of his most dire and ruthless foes to kill themselves.

For our current purposes—exploring the varied representations of the paranormal in psience fiction—there are three aspects of particular note in Wilson Tucker's somewhat innovative novel. Many other books by the mid-1950s had traced the early years of a telepath or other psi-gifted person, from the mutants Odd John and Jommy Cross to more conventional humans with an added quirk. Paul Breen's psi, like that of his fiancé and her brothers, is never explained and never really a point of interest to the telepaths—although the Neanderthal/Cro-Magnon analogy suggests some sort of selection pressures that yield a genomic alteration.

What is interesting here is watching the slow development of Breen's powers, and the studied, ruthless way they are exploited by government agencies. Is there much resemblance between this portrait of a cold war psychic spy and the operational and theoretical psi agents of Star Gate and the equivalent Soviet program? Almost certainly not. That doesn't mean black projects might not, then or now, have found and exploited super-psychics; we just can't yet know.

The second aspect is Tucker's assumption that telepathy is just like having a cell phone implanted behind your ear, or rather a device that can detect other people's thoughts at any distance and bring them to your awareness as if they were fully formed grammatical utterances. We are not shown the sort of blurting, blurry inner processes that might be expected by those who have listened to their own pre-verbal mutterings. Of course when two telepaths deliberately choose to converse by this non-vocal means, it might be reasonable to suppose that at some point the same kinds of grammatical structures will shape intentions into the equivalent of speech, and be literally heard in the way we hear voices in white noise or dreams.

Finally, Tucker's plotting implies that *Homo superior* (or *Teleman*, in the rather unfortunate coinage of Breen's fiancé's kindly and supportive *Homo sapiens* father) will finally always be able to escape even the most rigorous confinement, by whatever means necessary. Since presumably almost everyone reading this book is of the *H. sapiens sapiens* species, this might not be comforting news—although it might provide encouragement to treat outsiders with the same consideration that we give to those nearest and dearest to us.

14.
James H. Schmitz,
The Ties of Earth (1955)

Very different in tone from Schmitz's frolicsome *The Witches of Karres* is his grim short serial of the then-near future, begun in the November 1955 issue of *Galaxy*. Curiously, the second part of this two-parter is accompanied by an editorial by H.L. Gold who complains with *faux* bitterness about the general response of people to his own attitude to psi: "As far as I'm concerned, my motto is "Don't telepath—telegraph!" ... Psi is a nuisance. Every editor knows and dreads theme cycles, when writers all over simultaneously turn to the same idea. Right now, it's psi. I wish they'd pstop. I'm psick of psi" [*Galaxy* editorial, January 1956, p. 3].

This seems rather unkind toward Schmitz, and especially ironic since Gold would publish Alfred Bester's truly magisterial psi novel *The Stars My Destination* in just ten months' time. But it is an index of how thoroughly saturated the sf magazines were, by the middle of that decade, with psience fiction.

The representation of psi in this fragmentary, hodge-podge, jump-cut short novel is rather unlike its usual manifestations in psience fiction, especially of the 1950s' varieties. It seems almost like a manifesto from twenty or more years later—not so much New Age as Gaia-theoretic, presenting the

Earth as a single self-regulating organism with humans as its collective mind, at risk (in the novel) from the self-centered individualistic psionic New Minds arising from its ancient stock. In the real world, the Gaia hypothesis or paradigm was devised by Dr. James Lovelock in the late 1960s, and named for the mythic goddess of Earth. It would be borrowed in the late 1980s by Isaac Asimov for the final volumes of his extended *Foundation* sequence. In Schmitz's much earlier version, the secret mistress of war against the New Minds, who are represented by protagonist Alan Commager, proves to be his brainwashed, psi-gifted mother—Mrs. Lovelock. One is tempted to wonder if this was ironic precognition by Schmitz...

Commager is a tall, physically powerful fellow of thirty-four, proprietor of a business specializing in the care and sale of tropical fish. He is still recovering from the death of his wife, Lona, something of a metaphysics groupie who had hung out with their wealthy friend Ira Bohart, another cult junkie, and Ira's sporty wife, Jean, who shares Alan's skepticism. After winning big at a casino, Commager rather grudgingly attends a meeting of "the Guides" at the mansion of wealthy car dealer Herbert Hawkes. There he meets Ruth MacDonald, secretary of the Parapsychological Group of Long Beach, who seems rather too chummy with Ira, and Paylar, a woman in her early twenties, "downright cute in a slender, dark way ... another real personality" (Part 1, pp. 8–9). He agrees to take part in an ESP experiment, or half-mystical parlor game.

Then he is in his tropical fish store on Wilshire Boulevard, five hours later. He struggles to recall what happened. Paylar had asked him where his essence was. Commager indicated his forehead. More specifically? Between his ears, behind his eyes. "As a point of awareness, he seemed to be located an inch inside the right side of his skull. Simultaneously ... he noticed that his left ear was less than an inch and a half away from the same spot" (p. 12). She moves his awareness twenty feet in front of him, then twenty feet behind his head, swinging in space. He recalls no more, until he wanders in the store and finds the dead body of Ruth MacDonald. Huggermugger and confusion ensue. He takes the corpse back to the Hawkes mansion.

Seeking an explanation for these unusual events, he consults an old friend, a psychologist who cared for him as a young orphan and notes that these exteriorization "gimmicks" were got from "another group" (p. 19).[1] A hypnotist reveals that Alan has never been married; there was no Lona. That false memory quickly disperses, and Alan understands that it was based on the lovely Jean. Paylar informs him that the Guides wish to block the development of psi phenomena, and leaves in a car driven by Ruth MacDonald, conspicuously not dead after all. Curiouser and curiouser.

Commager is called to the tawdry fake-religious cult run by Dr. Wilson

Knox, who is terrified that the Guides mean to murder him; he is soon dead. Knox's assistant is a monstrously obese nurse, Mrs. Lovelock, who much later proves to be the true head of the antipsi forces, and Commager's mother. His late father was drowned by young Alan who understood the danger he was in from these Gaian defenders. All this is disclosed in gappy piecemeal episodes of attack and defense, building to the final exposition of how and why psi operates in the world.

Earth, Paylar reveals, is a single global entity in harmony with humans of the Old Mind. Evolved newcomers, the New Minds, lost touch with this unity, but were kept on the straight and narrow by Old Mind humans until New Minders

> developed a conscious interest in what was now called psi. That was an Earth life ability which had its purpose in keeping the patterns intact; and only the New Mind, which had intelligence without responsibility, was capable of using psi individualistically and destructively.
>
> In most periods of time, the New Mind was kept from investigating psi seriously by its superstitious dread of phenomena it couldn't rationalize—[Part 2, pp. 84–85].

The climax is an inevitable psychic shoot-out, but not one intended to be fatal; the Old Minders need subservient or brainwashed New Minders to contain and control the growing surge of New Mind psi. In a series of reversals that would stand the sf writer Keith Laumer in good stead a decade and more later, Paylar traps Commager by conditioning him to obedience and forgetfulness. This, of course, is the key to his plan: "the identity of Alan Commager was no longer absorbed by the consciousness that rose from and operated through his brain and body." He explains "gently" to Paylar what he has done and why. "For a few seconds, he encountered terror and resistance, but resignation came then, and finally understanding and a kind of contentment" (Part 2, p. 97).

Is this utterly horrifying, a perspective on psi that might make any ethical human turned away in revulsion, and perhaps do everything possible to exterminate this threat? Or should we, rather, rejoice as the psi-gifted "step into the role for which they had been evolved—and which the lower mind had been utterly unable to comprehend. To act as the matured new consciousness of the giant Earth-life organism."

This is, essentially, the same teleological proposition enacted in the closing pages of *More Than Human* and *To Marry Medusa*, and other psience fiction works that would follow. It has its lures and appeals, especially to brilliant outsiders ("Fans are Slans!"), but also to the duped multitudes of transcendentalist cults and faiths—and perhaps, too, of science itself. It is sheer power fantasy wedded to a redemptive, soteriological cover story. Just as well, really, that *The Ties of Earth* is, as scoffers like to say, "mere science fiction." But for all we know, perhaps it is not ... or might not always be.

15.
John Wyndham, *The Chrysalids/Re-Birth* (1955)

British novelist John Wyndham's important and classic novel of post-nuclear life is not quite psience fiction, although uncanny powers of the mind play a crucial part in the narrative of mutations induced by atomic radiation and fallout. In the mid–1950s, nuclear holocaust was a vivid and daily dread, not quite an expectation but nearly that. Wyndham's future, following alleged divine Tribulation for our wickedness, was not as uncompromisingly horrific and desolate as the ruined world of Cormac McCarthy's *The Road* (2006), which in any case probably resulted from an asteroid smashing into the planet rather than World War III.[1] Any plants, animals and humans that vary from a somewhat mythical pure form are quickly and piously destroyed, or banished to the wilds of the Fringes after sterilization.

So narrator David Strorm has to be cautious just because he is left-handed. That he has dreams of a wonderful technologically advanced city of horseless carts, flying carriages and lighted nights is something he has to keep to himself, as we learn on the opening page.[2] Is this is a psychic gift, a kind of reverse precognition, probably triggered by modified genes?

This suspicion is borne out when we learn that David, then aged 10, and some others in his Labrador farming community are telepaths. His younger sister Petra develops unusually powerful telepathic range and strength, and is the occasion of their discovery when she mentally controls a wild feline that attacks her horse. Their father, as so often in psi fiction, is a brutally self-righteous fundamentalist, determined to keep his family free of genetic "blasphemies."

When little David meets little Sophia from the far side of "the bank," an old freeway still in good shape from before the Tribulation, he learns from an accidentally unshod foot as they play that she has a tiny extra toe, and presumably one on the other foot as well. He cannot believe that this is enough to condemn her as a Blasphemy "hateful in the sight of God" (p. 14). The cruel certainties of his given world have received their first shake.

Unlike van Vogt's mutant slans, then, the chrysalids are a living reminder of a great nature-wide calamity. It is not only hidden telepathic abilities that fling David and the others into danger from scrupulously "normal" people, but any oddity of phenotype or mind. This is bigotry motivated by genuine fear of what horrors humans (or their deity) have unleashed on the world, ever ready to pop up from the loins of the apparently faithful and clean-living. The metaphor of the book remains as stark and valuable in the twenty-first century as in the middle of the twentieth.

And what if the psi portion of the terror were not merely a metaphor but a reality? Whether he meant it or not, Wyndham created a cautionary tale for those who proclaim the actuality of psychic phenomena. And we know from the experience of those involved in the U.S. government Star Gate program that a significant number of high ranking military decision-makers declared psi to be literally the work of Satan, a blasphemy, so that such research and operational projects should urgently be obliterated—as they were, after more than twenty years, in 1995.

What makes *The Chrysalids* notable as psience fiction is its engaging *ordinariness*, the sense that these imagined events are not utterly different from, say, life in a severely pious Amish-style community in which a handful of kids get hold of forbidden cell phones. Yes, David falls in love with his cousin Rosalind Morton, another telepath; they share a bond closer than most of us can manage unaided by easy psychic connectivity. But this intimacy might well follow, as many of us recall, from telephone calls so passionate and yearning that neither party can bear to hang up.

Even the especially potent psi range of Petra, which allows them to contact technologically advanced people from "Sealand" (perhaps New Zealand, less affected by bombs and fallout) and escape via airplane from persecution, is not intrinsically more marvelous than a short wave radio set and a handful of batteries. While Wyndham is admittedly a better stylist than many of the genre psience fiction writers, it is surely this *naturalization* of the paranormal that encouraged The New York Book Review to include the novel in their Classics reprint series a third of a century after first publication.

16.
Robert A. Heinlein, *Time for the Stars* (1956)

Special relativity was a fun topic for young readers in the mid-twentieth century, before quantum theory and then genomics became all the rage. It was built out of apparent paradoxes. It seemed to be true—leading experts swore by it, and would walk you through simplified versions of the apparent paradoxes—yet it seemed impossible to believe it, precisely because of those apparent paradoxes. Einstein and his supporters claimed that if you flew off in a very fast starship, accelerating to nearly the speed of light (the maximum speed possible in this universe), your clock and your very body would slow down by comparison to everyone you left behind on Earth. But wait—if everything is relative, and no standpoint is privileged, why not say that you on the starship are standing still and it's the Earth that's receding at light speed?

Wrong, it turned out. Why? Because the people on Earth never changed their frame of reference, whereas you and the starship crew did so every time you altered your acceleration. That proved it was you and not somehow everyone else back home who was subject to Einsteinian time dilation. You could rush off to a distant star a dozen light-years distance away, spend some time on a planet in orbit around it, and rush back home to arrive only a few years older—but find your unaccelerated family, friends and enemies a couple of *dozen* years older.

This bizarre effect had been incorporated into science fiction before Robert A. Heinlein built it into his novel that made literal this "twin paradox." L. Ron Hubbard's 1950 *Astounding* serial "To the Stars" was based on the consequences of time dilation, which mandated that star travelers would always return to a drastically altered Earth.

Heinlein wanted his travelers to remain in communication, despite their different time rates. Impossible, declared relativity, but then relativity did not (and still does not) allow for the observed fact that telepathy seems to operate instantaneously. Indeed, precognition suggests that information can travel faster than light, and hence backward in time. This implication is enough to make most scientists dismiss both paranormal phenomena as categorically impossible and hence not worth investigating. But science has been known to make errors in the past, and perhaps new physics will expand to allow for these observations.

Heinlein might not have believed this to be true, but for the purposes of his star-flight tale he allowed that consciousness might permit the simultaneity banned by relativity. Well, not really simultaneity, because he proposed that the Earthbound messages would seem to speed up when detected by the star-board, while the traveler's telepathic messages would seem to grind nearly to a stop when received back on Earth.

To get the full emotional bang out of this conjecture, Heinlein sent one identical twin into space while his brother remained at home. Tom and Pat Bartlett (one named for Thomas Paine and the other for Patrick Henry, American heroes admired by their father) were supernumeraries in the family, attracting a head tax in this overcrowded future and dooming all of them to cramped quarters. In search of extra income, they sign up with the Long Range Foundation whose genetics screening finds them suitable for an interstellar experiment in time dilated communication. They learn that while ten percent of twins from a single egg are telepathic, few others are (p. 29). One twin from each double who has tested successfully will travel to the stars on a hundred year exploration mission; the other will monitor interstellar transmissions that grow ever more sluggish in his or her frame of reference. Eventually, of course, the homebody will be too elderly for this task, and probably then die.

Pat, to Tom's annoyance, has always been a step ahead, top of the pecking order, even if nobody else can't tell them apart. So it is Pat who gets chosen for the trip. By a stroke of bad luck, at the last moment he breaks his legs skiing and Tom gets the chance of a lifetime. Of course Tom ends up in very dire circumstances when the starship *Lewis and Clarke* faces troubles on far worlds, then mutiny, and he understands that Pat was the true winner, safe at home, years older, finally with a wife and children, and then grandchildren.

For our purposes the adventures and lessons learned in the trajectory adolescence-to-maturity are not strictly important. Heinlein's careful thinking about duration-flexible telepathy is the psience fictional heart of the novel. What is it like, trying to communicate when your partner is eventually either droning too slowly to tolerate or blittering like a chipmunk? The experience is entirely subjective. You can't record the mental exchange. The grounding assumption, however, as often in psience fiction, is that psi transmits thoughts in a form that is received as grammatical speech rather than a Joycean stream of consciousness or a blizzard of visual images. This is perhaps allowable when exchanges are deliberate, at set times (as registered at either end), and therefore perhaps akin to rehearsing a speech without moving one's vocal organs.

From a Husserlian phenomenological point of view, perhaps the most valuable aspect of this novel is the emotional tenor of the telepathic give and take. The *Elsie* (the *Lewis and Clark*) has other telepairs, so Tom is allowed a day off while Pat has a spinal operation. While waiting, they take part in a calibration exercise, listening to ticking and matching them vocally. Of course the initially synchronized beeps fall out of identity; the time dilation effect is coming into play (p. 99). In the two weeks since launch, they have attained five percent of light speed, and the "slippage" in time rate has reached one point two parts per thousand. In a year they expect to be near light speed, and the local elapsing time on the ship will be close to zero compared with Earth.

And indeed by ship-time 13 months, twelve years have passed at home. Pat is married. Slippage has increased rapidly.

> At three quarters the speed of light, [Pat] complained that I was drawling [telepathically], while it seemed to me that he was starting to jabber. At nine-tenths the speed of light it was close to two for one, but we knew what was wrong now and I talked fast and he talked slow.

> At 99% of *c*, it was seven to one and all we could do to make ourselves understood. Later that day we fell out of touch entirely ... brains are flesh and blood, and thinking takes time ... and our time rates were out of gear.... It was upsetting [p. 139].

Less than a week later, but two years on Earth, Tom "hears" from Pat, under deep hypnosis and drugs to make him congruent with his brother. Not much later they begin deceleration and in two weeks they equal ten years on Earth—and his twin has a seven-year-old daughter. Several ship months later, little Molly gets a research grant to see if she can arrange a telepathic handover to her from Pat. It works. Not long after, there's a new baby girl as the *Elsie* reaches Tau Ceti, eleven light years from the solar system. It's a washout, a disaster. They accelerate again, past Beta Hydri. Pat is now 54, his wife has died, and Tom depends for his link on Earth to Molly and her daughter Kathleen, his great-niece. So luckily, and with no principled explanation, Heinlein assures us that the familial connection can pass down generations. That's not impossible; in the real world, children sometimes claim to be able to read the thoughts and feelings of distant grandparents, and vice versa, with some evidence that this is so.

Finally, poignantly, Tom is conversing with Vicky, his great-grand niece.

> Vicky grew up some when we peaked this last jump; now she takes notice of boys and does not have as much time for her ancient uncle.... This sounds as if I were jealous of a boy I'll never see over a girl I've never seen, but that is ridiculous. My interest is fatherly, or big-brotherly, even though I am effectively no relation to her [pp. 182–83].

There is a mutiny. Luckily, the data from the telepathy process has provided enough evidence supporting a variant of relativity that a new starship has been developed with an "irrelevant drive." It arrives a month later, and the *Elsie*'s crew are taken home at the same brisk rate. To the future, where Tom is a lost castaway: "the only news story I saw today was headed: THIRD LOAD OF RIP VAN WINKLES ARRIVE TODAY ... our clothes were quaint and our speech was quaint and we were all deliciously old-fashioned and a bit simple-minded" (pp. 233–34). Pat is eighty-nine and crotchety. Fortunately, Tom's telepathic link with little Vicky is still active, and even more fortunately Vicky is now old enough to tell him, mind to mind, "*After we are married, there will be none of this many-light-years-apart stuff.... You seem to forget that I have been reading your mind since I was a baby*" (p. 244).

If this sounds rather like little Goth's declaration of intent in *The Witches of Karres*, we must assume that something about psi and star travel is at work. Since we do not yet possess light-speed starships, though, there is not yet any way to assess the validity of these psience fictional precognitions.

17.
Frank M. Robinson, *The Power* (1956) and *Waiting* (1999)

Perhaps the most alarming narrative icon in psience fiction (and for some, if only secretly, the most desirable) is the mutant or superhuman able to take over an ordinary human's mind, creating a deluded puppet—as we saw at the end of *The Ties of Earth*. The trope is not restricted to psi stories, of course. Robert A. Heinlein's disturbing sf thriller *The Puppet Masters* (1952) was a paradigm instance: alien creatures had come from space in flying saucers and taken over the bodies and actions and even the self-awareness of ordinary citizens. The creatures were slimy things that settled on the back, inserting neural tendrils into the brain and nervous system. That was the old school method. In some texts at the margins of thriller, sf and explicit psience fiction, the capacity to enter another's ... call it *soul* ... from a distance is the driving force of the storyline.

Frank Robinson (or as he was known in sf fandom, Frqnk, from a 1940s fanzine typo) was a popular fiction writer in varied modes, such as ghost-writing the "Playboy Philosophy" purportedly by Hugh Hefner and the "Playboy Advisor" lifestyle column (Frank was gay). With Thomas N. Scortia he wrote *The Glass Inferno* (1974), later a basis for the blockbuster movie *The Towering Inferno*. Scortia was also his collaborator on *The Prometheus Crisis* (1975) and *The Nightmare Factor* (1978). He was a well-regarded sf writer, author of *The Dark Beyond the Stars* (1991), winner of a Lambda Literary Award given to novels celebrating LGBT themes. *The Power* falls somewhere in the midst of these thriller novels, exploring the experiences of a team of psychologists seeking ways to find or create superhuman soldiers—and realizing, as they die one after another, that one of these *Übermenchen* is hidden within their own company. It is a fertile idea for another movie, and was filmed by George Pal, starring George Hamilton, Michael Rennie and Suzanne Pleshette (1968), and earlier in a TV movie with Theodore Bikel.

Because the narrative looks over the shoulder of Bill Tanner, and indeed from within his own desperately frightened and questing mind, we know that the superman must be one of the other members of the team. Tanner's quest for this cruel, manipulative monster quickly leads him to the small town where the foe began his supernormal career as "Adam Hart," whom everyone in town recalls with loving fondness, even though as a teen he initiated brawls, got many girls pregnant then walked away from them, took money without repaying it. Each person recalling Hart describes him differently, in terms of their own ideal self-image; there are no unblurred photos of Adam Hart to be found. He is, in fact, the perfect would-be popular dictator. Is it mere chance that Robinson chooses his name for its resemblance to Ad[olf] H[i]t[ler], as well as intimations of the inauguration of a new man? But Hart is far more cunning and capable than the German hysteric. And he can, it seems, command other minds by his own unvarnished will. He is a kind of darkly Nietzchean anti-hero, and William Tanner has to stop him.

It is never quite clear if Hart is a telepath in the sense made familiar by *The Demolished Man* and other key psience fiction novels, but he can read and influence minds in some degree, although usually in a rather comic-strip way:

you don't think you're going to get away, do you, animal? don't run (p. 216)

His power does bear a certain resemblance to the final pre–Demolition scene in Bester's great book, where Ben Reich has the reality of his world ablated and erased item by monstrous item: "There are no stars." Hart has the power to make people forget Tanner's academic credentials and publications, leading to his firing from his professorial chair, and to remove all the evidence of his bank account and very history among some of his friends. How his physical thesis is caused to vanish from the university library, let alone his bankbook to be assigned retrospectively to someone else, is never even hand-wavingly explained. Hart has the Power, and that's all we need to know, which finally makes this novel a thriller rather than genuine psience fiction.

Still, it is a contribution to the nightmarish domain of "superhumans among us," and the denouement, when Bill Tanner is unmasked even to himself as another of the same kind, is both unnerving in a thriller way but also provocative for subsequent sf. The same device had been used in A.E. van Vogt's remarkable "space vampire" story "Asylum" (1942), and its extended version *Supermind* (1977), when a normal human reveals to himself that he is actually a masked Galactic with stupendously advanced IQ and powers, and in Damon Knight's serial "The Tree of Time"/*Beyond the Barrier* (1963/ 1964), where a professor of temporal physics learns that he is actually a Shefth, a demonic/angelic warrior from the far future. Neither of those works is strictly psience fictional, reinforcing our understanding that such literary categories are never watertight and in the final analysis are useful mainly as an index for readers who wish to categorize narratives bearing a general Wittgensteinian family resemblance.

Ψ

A psi-inflected reprise, although not a sequel to *The Power*, Robinson's *Waiting* (1999) is set some twenty years after the end of the Vietnam War,

The Power (1956) & Waiting (1999) 79

dating it at about 1995. Artie Banks, long since returned from the battlefield and three months in a Cong prison, now a TV news writer, identifies the mauled body of his medico friend. In the opening scene we watch Dr. Larry Shea running through the less savory districts of San Francisco, in terror, not seeing his pursuers, *herded*. He is set upon as voices whisper inside his head, and his face is gnawed away.

Three friendly, well-kept dogs are later taken in charge by the police, but we fantastika readers suspect that the murderers were some kind of hidden apex predator, maybe analogs of slans able to possess and direct the dogs. These suspicions firm when we learn that Dr. Shea was in possession of DNA and tissue samples from an elderly man killed in a brutal car crash; the dead man's organs, muscles and bones are those of a strong 30 year old, despite his apparent age. It seems likely that Shea was preparing a paper, and a talk, on this find. Warned, his wife and children have fled, and his house was subtly burgled. DNA samples and electronic records have vanished.

This murder and mind-control is not unlike the activities of the mutant superman in *The Power*, but Adam Hart was a freak, all but alone. This killer or killers, we shall learn, are more akin to Zenna Henderson's People, living in small cells or communities. But first Artie, like Shea and his work colleague Connie, is subjected to mind invasion. His wife, Susan, has gone for Christmas to care for her ailing father; Artie is depressed, and an inner voice tempts him to fall to his death. He is saved only by his 17-year-old son Mark, home in a wheelchair from his private academy. Now we know for certain that this is indeed psience fiction, of a sort.

Then, in a museum trialing a VR headset for kids, he hallucinates his role as an Old Human with simple language skills being set upon and eaten by the articulate Flat Face New ones. An anthropologist assures him dismissively that no descendant of Erectus or Neanderthalensis or any as yet unknown early hominin could still walk the planet. Artie knows better, having heard their inner whispers and shared this immersion in their "racial memory."

At length he knows that one at least of his longtime friends is of the Old People lineage, selected to mimic humankind. Do they read minds?

> "Don't be silly—nobody can read minds. We can't press a button on your mental computer and watch the words scroll up on a screen. We can see images in each other's minds and we can project them if you're receptive, that's all. It's like the Rhine experiments.... Many of your own children can do it when they're two years old. They mostly lose the ability.... When you're growing up, what's the one ability you always wished you had? To know what somebody else was thinking.... You lose that ability early, but you never get over wishing you had it back" [pp. 154–54].

The novel works diligently through the implications of this survival of a parallel hominin species. Several of Artie's nearest and dearest prove to be members of this clade, some willing to kill him, others working to save him. It is a disturbing if rather manipulative story, and the psi elements never resolve into a clear statement—except for Artie's realization that humanity is likely to be replaced by the Old People, and eventually they in turn by some other emergent people, just as the invincible and vastly numerous dinosaurs perished and tiny mammals took their place and then surpassed them.

If psi is a function of genetics and natural selection, perhaps that is indeed the inevitable future of our species, and finally theirs, if they exist already as a hidden and endogamous community.

18.
George O. Smith, *Highways in Hiding* (1956)

Here is a future that is routinely psi-facilitated, although as with programming or genetic engineering in our world there are different kinds of psi specialties, and of course (as in *The Demolished Man*), differing levels of competence. You can travel by coptercab or in your own turbine car on conventional roads and highways, and there are regions that are "dead zones" that inhibit psi, just as we have poor cell phone reception in certain locales. The Rhine Institute is held in high respect, and post-doctoral scholartes in psychic disciplines are especially favored.

A brief foreword tells us that this book is the first popular treatment to use the specialized punctuation denoting telepathic communication, which is what we'd call a hash mark, #. Psi comes in two major flavors: *espers* can "dig" the physical world with their power of *perception*, which is rather like an astral finger run over a surface or deep inside a physical body, and *telepaths*, who read minds. Espers use clairvoyance honed to a subtle skill, like the fingertips of the blind. (Those without esper are "blanks.") Telepathy is not possible with those lacking its gift, although their minds are open to those who possess this nonsensory quasi-sense.

Steve Cornell is an engineer who had planned a career as a D.Ing, a doctor of mechanical ingenuity, but he ran out of funds and ingenuity. He awakens in hospital after a brutal road crash that his esper perception should have spared him. His passenger and wife-to-be, Catherine Lewis, is nowhere to be found. Two men had discovered the wrecked vehicle and saved Cornell, but report no passenger, alive, maimed or dead, nor is his fiancée's bag found in the wreckage. A telepath physician, Thorndyke, confirms that Steve believes his own story, but attributes it to confabulation—while admitting that documents support the story. Police, many psi-gifted, get nowhere.

Cornell sets out to find his fiancée with no help from the authorities, starting with inquiries directed to the men who, he reported in his dazed post-rescue state, physically lifted his heavy vehicle off him like supermen. Phillip Harrison, one of his rescuers, confirms Catherine's absence at the wreck scene. Phil's sister Marian, an early twenties bombshell, shows up, and Steve, who earlier tested Phil's psi with a harsh mental observation, "made some complimentary but impolite mental observations about her figure, but Marian did not appear to notice. She was no telepath" (p. 26).

At home he falls into depression. "I wanted to project my mind out across some unknown space to reach for Catherine's mind. If we'd both been telepaths we could cross the universe to touch each other with that affectionate tenderness that mated telepaths always claim they have" (p. 30).

This introductory material deploys with some economy a considerable amount of background data about George O. Smith's psionic world while setting up the emotional basis for the remainder of the narrative. In his earlier conversation with the hospital telepath, a brief discussion reveals Thorndyke's interest in solving a horrifying disease, Mekstrom's, first identified two decades earlier, which progressively hardens the tissues of victims into a metallic substance until they die. This clue that the Harrisons are former Mekstrom's sufferers who have transformed into shock-proof physical superbeings is eventually confirmed by Steve. He discovers a rather Pynchonesque secret system of emblems on state highways[1] that direct those who know to the domiciles and factories of a hidden group, perhaps those who hold his lost Catherine. When he meets an adolescent girl on such a marked road and gives her a lift, she proves to be a telepath with a frighteningly hard arm and grip, apparently a case of correcting and enhancing Mekstrom's disease. What follows is quite a startlingly candid scene for the mid–1950s:

> She shut up like a clam when she realized that her mouthing had given me a chance to think, and I went into high gear with my perception: #Not bad—for a kid. Growing up fast. Been playing hooky from momma, leaving off your panties like the big girls do. I can tell by the elastic cord marks you had 'em on not long ago.#
>
> Seventeeners have a lot more modesty than they like to admit. She was stunned by my cold-blooded catalog of her body just long enough for me to make a quick lunge across her lap to the door handle on her side [p. 48].

Perhaps psychic molestation in defense of one's life and limb is no vice.

Ψ

George Smith's approach to fiction, like Wilson Tucker's, was not always entirely serious. When Steve pursues the secrets of Mekstrom's disease, he gets a cold response from researcher Dr. Lyon Sprague, "a tall man, as straight as a ramrod, with a firm jaw and a close-clipped moustache. He had an air like a thin-man's Captain Bligh. When he spoke, his voice was as clipped and precise as his moustache." Steve is told by a medical superior, Dr. Phelps, that "Dr. Sprague ... lacks the imagination and the sense of humor that makes a man brilliant in research." Put in his place, "Dr. Lyon Sprague decamped with alacrity." In reality, L[yon]. Sprague De Camp, who looked very much like Dr. Sprague, was one of the great imaginative humorists of mid-century science fiction and fantasy.

Steve hires a nurse, Miss Farrow, to act as his witness as he tracks the secret highway system. Exposition continues, usually with a veneer of relevance:

> "What's your telepath range? You've never told me."
> She replied instantly, "Intense concentration directed at me is about a half mile. Superficial thinking that might include me or my personality as a by-thought about five hundred yards. To pick up a thought that has nothing to do with me or my interests, not much more than a couple of hundred feet. Things that are definitely none of my business close down to forty or fifty feet."
> That was about the average for a person with a bit of psi training either in telepathy or in esper; it matched mine fairly well [p. 71].

Farrow disappears from a hotel where they stop to rest; Steve is informed, with many witnesses, that she was never there. This is a narrative reeking of paranoia or its simulation, a genre trope far from rare in psience fiction (especially the oeuvre of Philip K. Dick). But Smith presents the varieties of his imaginary psi in an interestingly novel way:

> Esper is not like eyesight, any more than you can hear printed words or perhaps carry on a conversation by watching the wiggly green line on an oscilloscope. I wished it was. Instead, esper gives you a grasp of materials and shapes and things in position with regard to other things. It is sort of like seeing something simultaneously from all sides, if you can imagine such a sensation. So instead of being able to esper-read the journal, I had to take it letter by letter by digging the shape of the ink on the page with respect to the paper and the other letters [p. 92].

Less novel, and somewhat troubling, is a recurrence of what is becoming a favorite trope in the 1950s: people with paranormal powers using them to modify or replace the memories and attitudes of others. A third of the way through Steve Cornell's odyssey, he learns not only that Mekstrom's disease can be cured, which results in people of incredible strength and probably longevity, but also that false memories were implanted in his brain as he lay in hospital after the car crash from which his beloved Catherine vanished. The same device, we recall, featured importantly in *The Ties of Earth* and *The Power*, and of course at the climax of *The Demolished Man* when Ben

Reich's very understanding of reality is stripped away from him until his consciousness is gone forever, leaving a blank slate awaiting a rewritten substitute self. Was this a reflection of the mid-century dread of unpatriotic subversion and "Communist brain-washing" (as in Richard Condon's 1959 novel *The Manchurian Candidate*, for example) and in fact in the CIA's MK-Ultra experiments between 1953 and 1966—although those were not widely exposed until 1975, in the Church Committee hearings?

Steve's travails in the remainder of the novel add no fresh insights into a possible psi package of uncanny abilities, so we shall leave them there for the reader to pursue. A critical question, as always with these psience fictions, is this: how plausible are such fictional powers? Are they more realistic than superstitions of malign witches, or infestations by demons, or levitating saints?

But even posing the question that way might abandon too much too soon to the scoffers. Can we be certain, for example, that humans actually can never cause an object to lift off the table driven by no propellant other than a mental intention? Can metal curl and flex and melt in the fingertips of a spoon-bender, or is that nothing better than a childish parlor trick? Are there telepaths in the world, even if they perforce lose the trick of it in childhood, and others with a clairvoyant or prophetic gift?

It turns out that even parapsychologists are divided on these matters. It might be that there is a highway leading the curious and the patient to the truth about psi, but it is also conceivable that it remains a highway in hiding, perhaps destined to remain occluded.

19.
Alfred Bester,
The Stars My Destination
(1956–1957)

Alfred Bester's strategy was always to lead the reader a merry dance, not to say a danse macabre, to leap from concealment with shouts and firecrackers, to lurk and entice and disguise and … unmask! Explosion! Concussion! When he was in form, his pace, attack, payoff were exemplary, dazzling. Sf critic and writer Damon Knight noticed all this more than sixty years ago, when Bester was writing at the top of his form:

> Dazzlement and enchantment are Bester's methods. His stories never stand still for a moment; they're forever tilting into motion, veering, doubling back, firing

off rockets to distract you.... Bester's science is all wrong, his characters are not characters but funny hats; but you never notice: he fires off a smoke-bomb, climbs a ladder, leaps from a trapeze, plays three bars of "God Save the King," swallows a sword and dives into three inches of water. Good heavens, what more do you want? [Knight (1956), 1967, p. 234].

Bester's brain sizzled with lunacy, knowing, cynical but flushed with a baroque unashamed romanticism that was not all that common under the gray banner of the 1950s. *The Demolished Man* (1953), discussed above, and *The Stars My Destination* (1956–57) were unforgettable neon poetry blazing against the suburban night. Bester's masterworks left one in no doubt of the protagonists' names. The Penguin edition back jacket blurb caught it with vulgar precision: "What is Gully Foyle? ... Saviour, liar, lecher, ghoul, walking cancer ... a man possessed ... a blazing hero of a science fiction novel that transcends its category."

This last claim, however, is precisely wrong, for the book is a quintessence that exactly epitomizes its genre category. Samuel R. Delany, who rightly esteems it, noted that "*The Stars My Destination* (or *Tiger! Tiger!* in its original title) is considered by many readers and writers, both in and outside the field, to be the greatest single sf novel.... It chronicles a social education, but within a society which, from our point of view, has gone mad" (Delany, 1978, 35).

More than that, it is the apogee of Bester's consistent struggles with a single theme: a psi cosmos under pressure, the heightened image of a compulsively driven individual bursting through the prison bars of nature and nurture both. In a world of teleporting "jaunters," Foyle is marked by demonic and transcendent stigmata: a blood-red tiger tattoo across his face, a calligraphy of silvery lines on his body where a cyborg operation has made him a powered superman. Reich and Foyle manifest as a Bergsonian evolutionary emergent embodied in one passionate, driven creature who hurtles through a world stripped to hard, brilliant, teleological metaphors, where wishing and sheer grit really can make it so.

Bester's crucial notion, now long abandoned by practicing biologists and philosophers, is that Nature is in some sense a Designer with a Plan and a Purpose, shaking the bottle of elan vital until it seethes and spurts. His novels (the later ones regrettably dreadful) are overgrown with grotesque coincidence, lucky accidents of history that have the obvious narrative merit of advancing the story with maximum attack but through their failure to offend us conveying as well, and more importantly, a subterranean awareness that in these universes Nature is a participant, a sort of psychic partisan, rooting for the seed-bearers.

The typical Besterian seed-bearers comprise a dyad. As we have seen above, *The Demolished Man* evoked that dyad again and again in Ben Reich/ Lincoln Powell (criminal/detective), Ben Reich/Craye D'Courtney (upstart/

tycoon, and son/father), Ben Reich/The Man With No Face (conscious/ unconscious selves). All of those were subsumed, quite deliberately on Bester's part, into an archetypal mandala of contest which can be represented as Eros/ Thanatos, Life/Death.

In *The Stars My Destination*, the dyad is above all Gully Foyle (passive lowest common denominator transformed, like the Count of Monte Cristo, into a vengeful hero when his wrecked spacecraft is deliberately bypassed by the Presteign *Vorga*) and Olivia Presteign (beautiful blind albino slaver who sees only in frequencies we cannot register except with machines—insufficient information to allow her to jaunte—and murderous mistress of the *Vorga*). Each is at once the other's sibling Other and Self. This is true at least in terms of narrative impulse, but the dialectic between them points to something grandiose and, in individual terms, almost unspecifiable: perhaps the emergent salient of Life itself, set against the frigid, uncaring vacuum of space-time.

On the social level, the ground halfway between the psychological rampaging of individual compulsion and the final magisterial epiphany of Foyle-as-a-god, the dyad is manifest as common humanity versus power elite. Foyle effects a one-man revolution in human consciousness and power by dispersing PyrE, a kind of primordial pre–Big Bang element, to the brutalized masses of the worlds. PyrE is the primal stuff of the universe, latent force in its purest form, responsive only to Will and Idea. On the one hand, Foyle's act seems precisely an unwitting metaphor for mid-fifties liberal aspiration. On the other, it is an intriguing figure (no doubt overdetermined) for the devastating potential of both art and science in the conduct and context of human affairs.[1]

All of this is conveyed without pretension in the usual Besterian helter-skelter pyrotechnics, inventive setpieces, and variants of French poet Guillaume Apollinaire's *calligrammes*, images and sounds and mental manifestations displayed on the page in concrete poetry forms, first trialed in *The Demolished Man* to represent the experience of telepathy and here as Foyle's transcendental or numinous connections to past, present and future.

Ψ

As a high point of the psience fiction sub-genre, *The Stars My Destination* is impressive despite the paucity of traditional or new icons of consciousness-in-action. The primary psi power in the novel is teleportation—the willed instantaneous transmission of one's body and appurtenances across space to a mind-focused destination never more than 1000 miles distant. Surprisingly enough, *New Worlds*' Leslie Flood was quite mistaken in claiming that Bester "pack[ed] into the story practically every device known to 'psience-fiction,' plus a few original twists of his own." One superbly effective novelty, often minimized in reprints after *Galaxy*'s serialization, is the synesthetic nature of Foyle's perception under threat. These visuals almost leap at the eye. When

> RED RECEDED FROM HIM and
> GREEN LIGHT ATTACKED and
> INDIGO UNDULATED WITH SICKENING SPEED LIKE A SHIVERING SNAKE and
> He was in a scintillating mist a snowflake cluster of stars a shower of liquid diamonds

these are meant to be illustrated as such on the page, as they were by artist Emsh in *Galaxy* but never adequately thereafter, along with other visual metaphors of jumbled perception. They are abnormal, even supernormal, but they are not paranormal in the psience fictional sense

Telepaths are rare, booked years in advance, but lovely Robin Wednesbury is disfigured, so to speak, by being born a Telesend, doomed to transmit her agitated thoughts to others, mind readers or not. On Mars, there is only a single telepath, the 70-year-old baby Sigurd Magsman. There's little if any sign of precognition, clairvoyance or mediumship, let alone Ben Reich's imposed Mass Cathexis.

Is it just exquisite memory or a kind of remote viewing that allows jaunters (those who learn to teleport) to visualize their destinations? In the Prologue that opens the novel, we are told:

> If you were in a dark room and unaware of where you were, it would be impossible to jaunte anywhere with safety. And if you knew where you were but intended to jaunte to a place you had never seen, you would never arrive alive. One cannot jaunte from an unknown departure point to an unknown destination. Both must be known, memorized and visualized [Berkley edition, p. 6].[2]

How is it, then, that brutal gangs of thieves—Jack jaunters—tear around the world, often to places beyond oceans where they could never have reached by plane or ship. How is it that certain high-powered professionals can appear inside the most resolutely screened and secret labyrinths and Star Chambers without ever having visited these inner sanctums? Is this just Bester losing track of his narrative frame, or is he hinting that what we understand about psi is never complete?

Foyle is forced to take this gift to a new expansive level: from abandonment in deep space to a damaged spacecraft, through time to the past and future, and finally from Earth to view nearby stars from the frigid depths of interstellar space: Rigel in Orion, Vega in Lyra, Aldeberan in Taurus, Antares in Scorpius, and at last back to the ruined spacecraft *Nomad*, now incorporated into the twisted corridors of a Trojan asteroid, home of the tattooed lunatic Scientific People. There, like a fetus in the womb, he awaits his apotheosis as the first human to Space Jaunte.

The preposterous and arbitrary physics and sociology of Bester's most brilliant novel do little to damage his reputation as one of the most exciting and literate sf writers. These lapses do, however, reduce the value of the novels regarded in their psience fiction avatar. A curious aspect of science fiction in light of psi, and vice versa, is that many of the most detailed and at least mildly plausible works are less well crafted than those at the peak of achievement. We shall examine more of these presently.

20.
Lan Wright,
A Man Called Destiny (1958)

In December 1958, *New Worlds*' editor John Carnell was enthusiastic in welcoming Lan Wright's latest serial, *A Man Called Destiny*: "the fundamental principles of the story are rooted in the psi powers, which we have not touched deeply upon since Wilson Tucker's outstanding serial "Wild Talent" in 1954. More than telepathy is involved this time, however—telekinesis, pyrotics, precognition ... and something else" [Part 1, p.4].

The added ingredient, it turned out, was imperviousness to physical attack. In an adventure story, this is both a great benefit for a protagonist, akin to Superman's invulnerability to anything other than some variety of Kryptonite, and a drawback in its tendency to preclude narrative tension.

Still, it raises the necessary evolutionary "Red Queen's Race" question (in which species have to run as fast as they can in order to stay where they are). If a bug is easy prey to an insectivore bird, how can its species escape being eaten to the last specimen? Often, it turns out, by random variations in its offspring, some of which might have a slight resemblance to a leaf or even to a larger competitive bird (the fake eyes on a moth's wings, say), or a nasty taste or smell. Such slightly different critters will survive, and so it goes in the race for persistence over the generations. Won't a psi-gifted ecology, therefore, also tend to evolve resistance to psi, or psychic invisibility, or other countervailing factors?

Richard Argyle, rather like Gully Foyle, is a bulky, tall, second engineer. The scrubby cargo starship *Lady Dawn* is stuck on dreary Jones' Planet—a world without a single telepath—awaiting a new drive to install. He's approached in a bar by one Spiros who has pursued him from lavish Rigel Five with news that Angela, his runaway wife, has been dead for six years. Before that she worked at a senior level with interstellar Trader tycoon Pietro Dellora, one of the richest men in the Galaxy, and wished to compensate

Argyle for her disappearance with a good job on Dellora Planet. Spiros will wait for him at the Hotel Galactica on Rigel Five, settled five centuries earlier, with its native population thoroughly under the human thumb. By the time Argyle arrives, Spiros has been dead for three days, shot by his own weapon in a locked room, apparently by a teleport (but why not by a "kineticist"?)—although the Galaxy has never known a teleport until now (Part 1, p. 17).

So the psience fictional credentials of the novel are established early. Some psi abilities are clearly widely known and accepted, although some evidently are just conjectured.

> Over the years telepaths had grown in stature from variety theater entertainers with a limited range, to a group of people whose importance to the Galactic community could not be measured in terms of mere physical wealth. On them had been placed the responsibility of maintaining rapid interstellar communication, for the bridge between the stars was impassable without their special gifts. Others had come to join them; kineticists...; prognosticators...; pyrotics, levitators and a dozen others. As mankind spread through the Galaxy his mind expanded and gave forth its power. The process was slow; so slow that almost no one noticed it except the scientists who probed and picked at the reasons behind the development [Part 1, pp. 37–38].

Spiros is identified as a telepath, and Angela, who foresaw her own death, as precognitive. Pietro, immensely obese, is obliged by his grotesque mass to live under reduced gravity in an orbital satellite and, without his prophetic assistant, is soon assassinated. His son Alfredo, the murderer, is recognized by Argyle as the teleport, after Argyle is framed for Pietro's murder. Armadeus "Preacher" Judd, President of the Terran Grand Council, is somehow involved in this. Additional deaths are blamed on Argyle. Galactic law and order are under threat by a conglomerate of Traders who hold the key to telepathic (or teepee) communication. How they do this is never spelled out.

What links this psi novel to a long list of sf-inflected crime and politics narratives, and to the paranoid aspirational vector of many "savior of the cosmos" sf stories, is the gradual revelation of Richard Argyle's role in his world's future. He is a dark horse of another color, a non-psi mutant with a prospect of living at least three hundred years, and he cannot be killed.

How does this work? It is explained thus:

> The Galaxy is aswarm with novelty.... The mind of Man giving up it's [sic] secrets, Argyle, because Man himself is being forced to adapt ... to new and ever-changing conditions.... [A]s our world expanded around us, so the mind of Man expanded too. We stretched our wings out across the deeps of space and reached the stars, and with the changing status of our race ... adapt[ed] to the everchanging needs of the universe. We accept the teepees for what they are—the means by which we communicate across the vastness of space between the stars.... In every generation there are more people born with odd gifts that show the adaptability of the species to new and strange conditions [Part 3, p. 98].

Argyle's role is to lead the other mutants to galactic domination, in the nicest possible way. Chief Minister Judd, like *Slan*'s secret mutant leader Kier Gray, is head of a covert organization of mutants with hearts of gold, and himself proves to be a powerful teepee. Judd's devoted servant Janus—"a little fellow with the face of a Pekinese" (Part 3, p. 89)—is more, a telepath with the power to alter and shape the attitudes of those he targets (a now-familiar trope that will recur importantly in Brunner's *Telepathist*). But Argyle, who turns out to have inherited command of Pietro's commercial kingdom and has the documentation to prove it, is their trump card. Judd declares:

> I can only stand by—once I have created the opening—and watch you take advantage of it—if you can. On what happens in the next few hours depends the fate of the Galaxy for the next thousand years.... The history of the Universe will be shaped by your hands, and yours alone. Don't forget it, Argyle, don't forget it for a single instant [Part 3, p. 110].

Can a man whose only skill is the inability to be killed succeed in snatching the galactic reins of power? Is cosmic teleology supporting him, or is blind natural selection at work? Gully Foyle would be literally on fire by this point, hurling himself blindly back and forth in time and space. Argyle's tale is less hysterically driven than that, but as the cutting edge of emergent psi forces in the narrative's cosmos he is destined (as the title assures us) of winning on the largest available scale. This happens, but in a rather disturbing way. Confronting the assembled grandees of the Trader corporations, backed by Judd and Janus, he hears a quite different account of the galaxy's destiny.

> In the beginning there were riots against these—these freaks, but later came tolerance and acceptance.... [The psi gifted then] became something more than just people. They became a mass, a living entity with power and ideas and ambitions.... They grew together, first for protection against those who would destroy them and then because they could find sympathy only with others of their own kind. Later came ambition. Much later came Armadeus Judd [Part 3, p. 117].

Judd and Janus represent the leading edge of what Theodore Sturgeon might have agreed was a kind of gestalt mentality, perhaps the kind found in *To Marry Medusa*, but far less ecumenical and species-embracing. This is not dog eat dog, as throughout history, but more akin to Frank Robinson's Old People versus New People, genus against genus, perhaps ultimately conflicting murderous species. The son of Alfredo, claiming control of the Company Dellora, is physically choked by Argyle until his only recourse is to teleport away in full view of his wealthy and powerful human plotters, to their dismay. When he returns with a weapon aimed at Argyle, he is shot to death by one of the grandees, briefly a puppet under the control of Janus. This is presented as a triumph for the future of the universe: "They could only live with hope that they would be successful and with the unconquerable knowledge that they were right" (Part 3, p. 124). To a disinterested observer, this might seem the self-justifying conviction of the founder of a thousand-year psi tyranny.

21.
Marion Zimmer Bradley, *Darkover* Series (1958–)

Just as some novels hover at the border between science fiction and thriller (Frank Robinson's *The Power*, for example), many nowadays sprawl between sf and fantasy or alternative histories (the multitude of "steampunk" novels, say, or such dragon-inflected allohistories as Naomi Novik's *Temeraire* sequence).

Perhaps the most influential of the second complex category above are the very numerous Darkover novels from Marion Bradley (1930–1999), many co-written or entirely composed by others after her death, and the several chains of linked fiction by Anne McCaffrey (1926–2011), notably the Dragonriders of Pern series with some 24 volumes.

Both of these aggregations have backgrounds that seem typically science fictional—another world, in an interstellar setting of reachable stars and inhabited planets—but modified in the direction of fantasy by the psychic powers featuring prominently in both. (Dragons are not themselves fantastical, just unlikely—there were pteranodons, after all, with twenty-foot wingspan. Psi abilities are not necessarily fantastical either, unless they are treated that way, as they often are in the marketing of such novels.) In this sense, then, they are highly significant, in their sheer bulk and large devoted readership, to the psience fiction sub-genre.

Bradley's corpus, begun in 1958 with *The Planet Savers*, continued for at least another 40 volumes by her and her co-authors.[1] By the relentlessly fecund standards of the sharecrop bibliographies of the *Star Trek* and *Star Wars* transcriptions and *hommages*, this is not remarkable—indeed, for a genre where novelty and ingenuity were once prized, it is increasingly usual. (Probably in part we can blame, or thank, those Golden Age masters of ingenious novelty, Robert A. Heinlein, Isaac Asimov, and Frank Herbert.) For that reason, and because with these two large sequences it is the overall psi structure of the stories and novels that mark them as psience fiction, it will suffice to explore and annotate the psionic devices in play.

In the Darkover cosmos, which entails elaborate world-building by Bradley, the red giant star Cottman has a number of worlds but only one that is habitable, with four moons that create complicated weather and tides. Before humans arrived, long ago, Darkover's narrow viable strip was populated by an indigenous humanoid species, the Chieri, hermaphrodites who are highly developed in psi capabilities which can be passed on to the hybrid offspring of Chieri and human species. Other intelligent species coexist, including some resulting from genetic engineering.

When the air is warm (rarely, on this bleak world) a plant, kiraseth, releases a pollen that supercharges telepathic input in the sensitive and tends to drive people and animals mad. The varieties of psi skills and vulnerabilities are known generically as *laran*, chiefly telepathy and psychokinesis. Much of the plotting in these tales depends on mastery of laran, especially when assisted by *star matrixes* [sic, not *matrices*], a vividly luminous blue crystal or gem. What powers psi effects is *energons*, a kind of brain energy (like the traditional Chinese *chi* or Indian *prana*, perhaps, or the "subtle energies" of some parapsychologists) that flows through the body's meridian channels. Sexual function is also intimately connected to this energy flow, and the channels have to be kept in top condition to avoid dangerous illness, especially pubescent "threshold sickness." A star matrix can accumulate energons and enhance laran.

In the real world, this doesn't make much sense. If, as many psychics attest, large heavy objects can be lifted or flung by PK alone, the comparatively small amount of energy in the brain (about a fifth of the body's total energy consumption)[2] could not begin to power such an effect except if it worked as a gate or trigger to some mysterious outside forces as yet unknown.

It is worth bearing in mind, though, that to the extent such psience fiction draws on actual reported and well-witnessed anomalies, the attempted explanations offered by metaphysicians in the 18th and 19th centuries, and even today, need not be taken literally, nor the observations dismissed on the grounds that available, fanciful explanations are no longer plausible. Steam engines worked even though phlogiston (a hypothetical mass-less fluid flowing from hot objects) did not exist.

In any case, given this postulated laran psychic economy, the social and personal structures of life on Darkover differ from our own, and represents a deliberate attempt by Marion Zimmer Bradley to imagine a world with a history of using and increasing psi arcana. In a novel such as *The Demolished Man*, ESP is proved and accepted because it is so useful and gives power to its possessors, as electricity did in the 19th and later centuries—but there has not yet been sufficient time for psi to be utterly normal. What kind of culture, or clash of competing cultures, would emerge on a planet with people empowered by laran? Can we find sufficient parallels and lessons in the mythologies of world religions? Or do satisfactory psience fiction sagas require pioneering and unforeseen complications and opportunities?

Certainly the Darkover lineages and their associated psi gifts are numerous and somewhat specialized. In the Comyn Families, psi manifestations range from forced mental rapport to seeing someone as they were or shall be in their past or future; Foresight or general precognition; Catalyst telepathy that awakens the laran in those in whom it is latent, although risking illness; empathy even with animals; sensing and commanding weather patterns (something alleged of the late Ted Owens, the so-called PK Man)[3]; and so on.

Evidently in playful mood during this epic world-building, Bradley named some of these families in a jesting Tuckerization: the Dellrays, who also know which way the weather is turning, are plainly derived from Lester and Judy-Lyn del Rey, publishers of many wildly popular fantasy trilogies in five or more fat volumes (*The Sword of Shannara*, for example, by Terry Brooks, their first great sub-Tolkienesque success), and the Deslucido, seemingly a tip of the hat to married TV comedic stars Desi Arnaz and Lucille Ball, although Bradley's Deslucido family is noted for their ability to deceive even in telepathy and are therefore destroyed, so maybe their name is from the Spanish word meaning shabby or worn out.

Short of reading all or even many of the volumes, a good guide to the world Darkover, its human families, aliens, and laran powers is available online at the Darkover Wiki, not to be confused with Wikipedia.[4]

22.
Jack Vance, "Parapsyche," "The Miracle Worker" and "Telek" (1958)

One of the most accomplished prosodists in the sf spectrum, Jack Vance often moved along the narrative margins between science fiction and fantasy, and was marvelous in both territories. Psience fiction often shares that boundary status, and it is fitting that Vance from his early days as a professional writer ventured into the paranormal with the conviction of a denizen. While none of his novels is a clear-cut example of psi-based sf, a number of his novellas are suitable for examination here. We should look at "Telek" (1952), "Parapsyche" (1958), and "The Miracle-Workers" (1958), with emphasis on the latter two more sophisticated stories.

Many of Vance's works are ingenious and baroque; Norman Spinrad notes that "Like a painter, he endlessly describes clothing, architecture, landscape, and qualities of light for the purely aesthetic joy of it.... Not content to limit himself to the mere world-creation of traditional science fiction, Vance adds those graceful superfluities that give his times and places baronial richness, late Renaissance grandeur, and the weight of cultural and aesthetic substantiality ... combined with Vance's characteristic sardonic viewpoint and his relentless sense of irony" (Spinrad 1980, pp. 16–17).

"Parapsyche," though, unusually for Vance with his dominant rich and inventive sf voice or range of voices, is a very plainly told naturalistic portrait of life in California around the time up to and during the Korean War, except that its two young married principals are fascinated by the paranormal and determined to understand how it works. It forms a good basis for grasping Vance's approach to unusual phenomena, sketched in his comment cited in the collection *Minding the Stars*: "I had been doing some reading in the field of psionics and decided to expatiate upon my own theories, using the story for the vehicle.... I can't pretend to offer enlightenment; there isn't any to be had."[1]

His fictional explorations of psi, he observed, embodied his fascination with the unknown. Although the whole range of purported psi capacities is in some measure explained by the hypothesis advanced and enacted in the novella, possible traces of life beyond death are the device giving rise to that somewhat Jungian hypothesis. Fifteen-year-old Jean Marsile and her seventeen-year-old boyfriend Don Berwick go with teenaged schoolmates to the burned-out husk of a house destroyed by a madman after a cruel murder. The youths are appalled by what seems to be a re-enactment of this murder, sighted through an empty window, experiencing "a sad lonely feeling, deep as a pit. A coldness—" Was the image of the burning woman a genuine ghost, or some kind of memory trace etched into the murder scene?

Jean's father Art is open-minded, but her fundamentalist (half?) brother Hugh furiously mocks the apparition as blasphemous. When Don decides to spend his life seeking answers to the mysteries of parapsychology, the increasingly fascistic and bigoted "Christian crusader" Hugh is set on a course that must lead inevitably to Don's death. But if Don's and Jean's joint researches are valid, perhaps it will lead also to Don's persistence beyond death or even his resurrection.

Before that denouement, Don gathers scientific associates in his quest for insight into the occult. Jung's collective unconscious is the reservoir of human symbols and ideas which he intends to expand, proposing that "the so-called after-life is identical to the collective unconscious of the human race."

This might seem a somewhat empty proposition. Where does this shared consciousness reside? In an overlapping cosmos with different dimensionality?

Or in traces shared in the physical brains of the living? Don wonders whether dowsing is related to ghosts? How can telepathy work? Is the future perceivable? Do ghosts somehow live and think? Are they merely imprints akin to footsteps? If they do live, what is it like there?

The validity of dowsing is demonstrated by Art Marsile's success in finding water two hundred feet below ground, and then his dowser locates oil very much deeper in an adjacent portion of their property. After Art dies in an accident, he speaks to the couple via a medium, telling them (with Vancian humor), "Well, we stopped too soon. I just kinda pushed my head down and took a look." Trusting this possible communication from the dead, they do so and become overnight millionaires.

Hugh is furious at losing his supposed share of the wealth, although he earlier derided their superstitious dowsing and abandoned his claim to that portion of the land, preventing him now from using it to fund his crusade for an America hailed in his posters as Clean, White and Christian. He vowed to fight Communism, Atheism and "Blood Pollution."

Nothing is left for the reader's assessment at the close. Hugh kills Don, and is ready to murder his sister as well. She shoots him dead. And we are taken with Don to the realm beyond death. The afterlife, he sees, is a construct, explaining the variant descriptions provided by a team of mediums. Time and space are equally flexible. Yet this is not a one-way street. Because the afterlife is part of the collective unconscious, the esteem with which the deceased are recalled modifies their postmortem status and power to act. Hugh is venerated as a messiah by his loathsome followers, so in the other world he is magnified and dangerous. Don's own eminence is multiplied when Jean speaks out for her slain husband. In a rather comic book contest, dead Don defeats his dead foe, then in an even more cinematic climax returns to earth and drives off with Jean into the night: "the tail-lights became a pair of red dots, a glimmer, and then were lost." We, however, are not lost: the secret of the afterlife has been not only explained but proved.

Ψ

In "Telek," published six years earlier, we are still on Earth but some century or two in the future (despite the many unorthodox familial and personal names, usually a signifier of far futurity). It is a world drastically altered by the emergence just sixty years earlier of quite ordinary humans with the power of psychokinesis (or as the story names it, "telekinesis," the old term from the days of mediums and spiritists). In a population of four billion, they are just 4000 in number, but their ability to move matter by simply looking and willing has utterly reshaped human culture. Angry people thrown out of work or their familial homes on the whim of a Telek have no recourse; a Telek slain in reprisal brings upon the neighborhood a fearsome vengeance, and the instruments of justice are without avail against these godlike and gaudy individuals.

Is a mass revolt feasible or desirable? Grouchy Geskamp and young firebrand Will Shorn, an architectural draughtsman, discuss the topic; Shorn means to plant explosives in the foundations of a grand new arena being constructed for the glorification of the world's assembled Teleks. When a complaining former farmer is flung into the air and smashed down to his death by a peeved Telek, Forence Nollinrude, Geskamp kills Nollinrude and Shorn hides the corpse in a concrete pour. Geskamp is shot by local law officers.

When Shorn's older associate Circumbright urges compromise, Will dismisses its feasibility:

> One time it was that way. The first generation. The Teleks were still common men, perhaps a little peculiar in that things always turned out lucky for them. Then Joffrey and his Telekinetic Congress, and the reinforcing, the catalysis, the forcing, whatever it was—and suddenly they're different.

What's needed is a living cooperative Telek who can be studied, so that the rest of humanity can learn the secrets of the paranormal. Shorn agrees to impersonate a young Telek, and manages to trade the secret threat to the mined arena for the secret of transformation. This occurs; it seems rather improbably to be a matter of summoning sufficient confidence, ideally in the company of Teleks who provide a boost to the neophyte. "Your perspective must be adjusted; you'll be living with a new orientation toward life," he is told. And indeed as the psychokinetic power flows into Shorn, he understands that he has incorporated the world into his sense of self, so that lifting a block of iron is as natural as raising his arm. Soon he is flying faster than light to the Moon, Mars, Venus, bringing back trophies.

The crisis is arranged at the 60th Telekinetic Conference, and carried off with only a few people necessarily killed. Hundreds of new Teleks loft into the air, transformed by the collective effect of so many Teleks creating what Alfred Bester a year earlier had epitomized as a Mass Cathexis. The insurgents spread this psychic conversion through the world. It is not remotely plausible, but it is undeniably psience fiction.

Ψ

In a strong essay on Vance's stylistics, Richard Tiedman comments that, like "Telek," "The Miracle Workers" is

> a study of ... the completely divergent paths which an isolated society may take if a small group is invested with extraordinary powers ... in "The Miracle Workers" ... the entire social system is based on their powers without which it would collapse ... demonstrat[ing] how purely imaginary chains can be used to bind a susceptible populace. Each of the [purported psi masters] has possession

of a demon, not in actuality, but in accepted belief—and what is not disputed soon becomes fact [1980, p. 209].

Within the novella's social structure, humans who landed sixteen hundred years earlier on Pangborn (a planet previously populated by the resentful First Folk) now engage in almost medieval warfare, spurning the "mystical" empiricism of their ancestors. Does this mean that the dread jinx ruling them, or contesting to do so, is entirely a nocebo, an imaginary force that creates harm only by suggestion? That seems inconsistent with much that is displayed. Inside his tent, the Head Jinxman Hein Huss thumps a doll and outside there is a gasp of pain from the apprentice Sam Salazar. Or perhaps Sam, a bumbling but curious youth, has simply memorized the sounds of afflictions visited upon his hex depiction. Is this novella, then, genuine psience fiction or just an emulation of that form, as the miracles wrought on Pangborn are merely tricks of a magician's trade?

Vance's exposition of this ambiguity is deft. A war party of Lord Faide's knights, foot soldiers, jinxmen and cabalmen leaves his Keep and approaches that of Lord Ballant. Their way is blocked by plantations grown by the autochthones. Faide seeks Hein Huss's prophecy and is told gruffly: "There are many futures. In certain of these futures you pass. In others you do not. I cannot ordain these futures" (p. 11). This is more candid than most of the evasions of classic "psychic" mountebanks. As Huss elaborates later (and here Vance foreshadows cautions applied decades later in remote viewing protocols):

> In the next instant there is only one future. A minute hence there are four futures.... A billion futures could not express all the possibilities of tomorrow. Of these billion, some are more probable than others. It is true that these probable futures sometimes send a delicate influence into the jinxman's brain. But unless he is completely impersonal and disinterested, his own desires overwhelm this influence [pp. 23–24].

At Ballant Keep, Lord Faide's forces and apparently magical demons (frightful masks, animated robotic beasts) bring him victory, but their way home is barred by the unemotional but canny First Folk. Even after using an ancient floating vehicle to set much of the forest alight, his forces are seriously reduced, with no magical intervention applied against the autochthones. ("We cannot hoodoo the First Folk," Huss explains straight-faced. "There are technical reasons.") Faide retreats to his own Keep.

Finally we learn why these masters of telepathy and modest teleportation cannot prevail against nonhumans. Huss breaks the traditions of his craft and explains to Faide that the miracle workers—those ancient humans who came to Pangborn in impossible machines—"used ideas and forces they knew to be imaginary and irrational" (p. 53). This might seem to the reader to be the delusion of those who have forgotten science, but perhaps it is justified in the light of quantum incomprehensibility. "We are rational and practical," he adds, teleporting a candelabrum several inches, "but we cannot achieve the effects of the ancient magicians." We might see an implicit distinction here with the almost godlike psi powers of the Teleks, who are able to nudge the planet Jupiter briefly out of its orbit and back again.

Jinxmen, Huss goes on, are experts in human psychology, knowing what drives people and the symbols that manipulate them. Since the First Folk are not cognate to this body of knowledge and practice, they are not vulnerable to "hoodoo." It is not that humans are tricked into believing in psi powers; rather, the process of hoodoo operates on conscious, unconscious and cellular levels. A jinxman learns to enter the aware mind and the deeper mind of his target, makes foe and hex doll one in his own mind, bringing his attack to bear by conveying its noxious message mentally. The victim then brings pain and doom upon himself.

All this requires merging minds, but alien beings do not have the same kinds of cellular structure as humans, or share sympathies. Can they be brought into congruence? Perhaps by entering the Wildwood and becoming one with the First Folk. As expected, the apprentice Sam Salazar wishes to undertake this task, in the company of the two leading Jinxmen. The three enter the alien realm of the First Folk and find extensive selective breeding of living weapons. They conclude that the Folk are not quite individuals but share a collective mind—and are told bluntly that the intent is to "kill men," who are unwelcome intruders, and return the planet to the moss. Their approach mimics the trial-and-error empirical scientific method of the earliest humans, and Huss painfully asks if the humans must also return to the irrationality of their ancestors, the miracle workers.

The story becomes a contest of two ways of thought, or three: the settled wisdom of the psychic, the unthinkable drives of the alien First Folk, and the pragmatic tinkering, entirely untheorized as yet, of Sam Salazar. When the Folk emerge from their forest and build a vast enclosure from the foam of their own bodies, Salazar of course stumbles on a household liquid that will dissolve the hardened foam. The future is set: science will return to Pangborn, and generations hence the new miracle-workers might build a new starcraft to return humans to their home world.

$$\Psi$$

This intriguing Vancian celebration of difference is, therefore, a sort of repudiation of psience fiction in the face of superior science fiction. Or, rather, in this instantiation at least, psience fiction merges with science fiction. It was the cover story on its appearance in Campbell's *Astounding* in July 1958, and in a curious way perhaps marks a transition from the obsession with psi.

From about this time, Campbell's infatuation with quirky wild or fringe science turns toward implausible machines and natural phenomena that seems to defy the known laws of physics and astronomy: the "Dean Drive," a supposed method for changing rotary motion into one-directional, reactionless thrust; a project to learn whether planetary astrology actually influences the weather; an argument in favor of slavery (not necessarily racially based). He had already abandoned his most derided enthusiasm from the dawn of the 1950s, Hubbard's Dianetics, and his promotion of the allegedly psionic "symbolic Hieronymus Machine," introduced in the mid-1950s, was ebbing into silence. Psi remained an element in some of his most liked choices—notable the two Frank Herbert serials that became *Dune*, and also Anne McCaffrey's dragonriders, Gordon Dickson's *The Pritcher Mass*—but he missed the seething cauldron of Philip K. Dick's dementedly inventive mind, and some other important psience fiction, as he had missed the thunderbolts from Alfred Bester.

John Wood Campbell died in 1971, aged just barely 61. John Holbrook "Jack" Vance died in 2013, blind, at the age of nearly 97; his last luminous, elegant novel, *Lurulu*, was published when he was 88.

23.
Short Stories (1949–1957)

Our focus so far has been held mostly on novels, where the ideas underlying psience fiction can be explored in greatest detail. Nonetheless, it is certainly important to remember that many of the gems of sf have been cut and polished to the smaller size of short stories (up to 7500 words), novelettes (7500 to 17,000), or novellas (17,000 to 40,000). For the short, sharp shock, these more compressed and shapely formats are often ideal. Fresh insights and "thought variants" are frequently best delivered in briefer work, and that was especially the case in the days when most science fiction was available in monthly or bimonthly magazines rather than hard cover or paperback books. Luckily the best of these stories has been preserved in anthologies or collections, although in what follows I shall sometimes reference the original publications.

Katherine MacLean, *"Defense Mechanism"* (1949)

Just as John W. Campbell was getting up a full head of steam for his long campaign in favor of psi in science fiction, and as a real phenomenon in the greater world, he wrote this cautionary header for the first published story by Katherine MacLean (1925–): "Telepathic power would be a wonderful gift—or is it? There might be some question as to whether it, and its possessor, could survive, after all" (p. 155).

Ted is a writer of light articles for popular magazines, looking after baby Jake while his wife, Martha, does the grocery shopping. He muses on feedback circuits and conditioning, now and then harmonizing with the infant's mind. Here in Connecticut "away from the mental blare of crowded places" (p. 156), "all the luminescent streaks of thought" can merge with "the calm meaningless ebb and flow of waves in the small sleeping mind." Martha returns and they start preparing a fillet meal as Jake wakes and wails. Ted translates his small messages, to Martha's disquiet; she feels that doing so "isn't *right*" (p. 158).

Ted "cheerfully" tries to explain why she's wrong: "E.S.P. is queer.... But what Jake has is just limited telepathy. It is starting out fuzzy and muddled and developing toward accuracy by plenty of trial and error, like sight, or any other normal sense" (p. 159).

This is rather a close approximation to the experiences of remote viewing and Ganzfeld telepathy or clairvoyance, as reported by research parapsychologists (although not nearly as much was widely known about such phenomena when MacLean wrote this piece of psience fiction). The disturbing extrapolation of the story is a kind of savage negative feedback that kicks in as a mental defense mechanism when Jake's favorite bunny is trapped by an out of season hunter and threatened with death. Ted confronts the hunter, heavy billet of wood in hand, and cuts the terrified rabbit's throat. "There was a crowded feeling in his head.... It was difficult to breathe, difficult to think" (p. 162). He lays the man out with a single clumsy blow. The hunter, he understands, is psychotic.

The reader, though, has no reason to agree with this estimate. The hunter was just doing what all animals, including humans, must do—what Martha, indeed, was doing at this very moment, preparing meat for their dinner. Life, by this reckoning, when grasped at the deepest level of shared awareness, is psychotic in its very nature, and unbearable unless the shutters are slammed shut. "And Jake lay awake in his pen.... He would be a normal baby, as Ted had been, and as Ted's father before him. And as all mankind was 'normal'" (p. 162).

C.M. Kornbluth, *"The Mindworm" (1950)*

One of the major tragedies of classic sf was the early death at 34, from cardiac arrest, of Cyril Kornbluth (1923–58). A brilliant sardonic and worldly wise New Yorker, he wrote copiously both alone and with Frederik Pohl and

Judith Merril. In 1950, "The Mindworm"[1] appeared in the very short-lived *Worlds Beyond*, edited Damon Knight. (In his critical study *In Search of Wonder*, Knight deplored the waste of Kornbluth's talents, which represented "the triumph of a master technician over an inappropriate form—as if, on a somewhat grander scale, Milton had written 'Paradise Lost' in limericks, and made you like it" (p. 149). Even so, "The Mindworm" conveys a disturbing sense of what telepathy might be like deployed by an orphan psychopath.

Part of the effectiveness of this story is Kornbluth's varied methods for displaying in text what cannot be conveyed in words, rather as Bester would do the following year in *The Demolished Man*:

> The Mindworm, drowsing, suddenly felt the sharp sting of danger. He cast out through the great city, dragging tentacles of thought:
> "...die if she don't let me..."
> "...six an' six is twelve an' carry one an' three is four..."
> "...gobblegobble madre de dios pero soy gobblegobble..."
> "...melt resin add the silver chloride and dissolve in oil of lavender" [p. 69].

Another psychically "overheard" stream of consciousness is not at all Joycean, and chillingly alien in its controlled intelligence:

> Some people were walking forward from the diner. One was thinking: "Different-looking fellow, (a) he's aberrant, (b) he's nonaberrant and ill. Cancel (b)—respiration normal, skin smooth and healthy, no tremor of limbs, well-groomed. Is aberrant (1) trivially. (2) significantly. Cancel (1)—displayed no involuntary interest when ... odd! *Running* for the washroom!" [p. 71].

The ending is unsatisfactory: the Mindworm proves to devour the vital energies of normal humans, but is at last tracked to his lair by "mustached old men with their shirt-sleeves rolled down" who shriek WAMPIR, stake his heart and scythe his throat. Still, as an exercise in the psience fictional delineation of mind powers, it was a notable early effort not without influence.

Walter M. Miller, Jr., *"Command Performance" (1952)*

Returned from the Second World War a broken man, probably with PTSD, Miller (1923–96) worked variously, wrote increasingly strong fiction in the 1950s. This important contribution to psience fiction shares some tonal qualities with Kornbluth's (another victim of war damage, although his was physical rather than emotional). It appears in his collections as "Anybody Else Like Me?" which is often the despairing cry of sf telepaths and other psions with no awareness of other people sharing their ambiguous gifts. When she finally does meet another, she is crushed by the consequences.

"Quiet misery in a darkened room" (*Galaxy*, p. 141). Young, beautiful mother of three Lisa Waverly suffers a "crawling of the mind," and "unfed hunger." She is smart and loves her "good dependable" husband, but when Frank leaves on business and her kids stay with her mother, Kenneth Grearly approaches her. He, like her, is a nascent telepath, but he understands their mutant gift, and is convinced (as often in psience fiction) that they have a moral obligation to have children together, to advance the evolutionary tendency toward mental communication. This, he is convinced, will solve many social and personal problems, bypassing the low-bandwidth restrictions of speech and writing. Happily married, Lisa is appalled by this brazen approach. He manipulates her perceptions, intent on rape. As he steps from a curb toward her house, she floods his mind in turn with confusing images, the precise opposite of how Grearly claims expansive psi should help bond people.

> She imagined a fire engine thundering toward her like a juggernaut, rumbling and wailing. She imagined another car racing out into the intersection, with herself caught in the cross-fire...
> A real car confused the scene...
> There was a moment of rending pain, and then the vision was gone [p. 159].

Grearly is dead. Now she is flung back into solitude, more alone and despairing than before.

> "*Is there anybody like me? Can anybody hear me?*"
> There was only complete silence, the silence of the voiceless void.
> And for the first time in her life she felt the confinement of total isolation and knew it for what it was [p. 160].

Isaac Asimov, *"Belief"* (1953)

Doctor Asimov (1920–92) was famously a skeptic on such topics as psi, and grew increasingly irritated by his great mentor Campbell's fringe enthusiasms. When he wrote a story about levitation, Campbell returned it—but not because of any disbelief expressed in the tale. He did not appreciate the downbeat ending. So Asimov altered it, grudgingly. Both the published and the offending portion of the original version can be read in his collection *The Alternate Asimovs* (1987). Thus the story or stories are a case study in psience fiction from an unusual angle.

"Belief" is a somewhat plodding whimsy (if that's possible), with occasional smiles. Really it's a gentle outburst (if that's possible) against academic rigidity, bias, inability to deviate from the reigning paradigm. In the original version the whimsy slowly darkens, and ends at dead black. This is a worthy topic, but it might be hard to take seriously the notion, as Asimov dramatizes it, that the professors would simple turn their backs, muttering derisively, and refuse to accept that a man can fly even as he hovers above them. Although many parapsychologists are familiar with exactly that kind of dismissal in the face of admittedly less dramatic evidence—because usually the published evidence for psi is rejected without being read, since many scientists just know in their highly trained bones that it can't exist, therefore the supposed evidence is lies or stupid incompetence.

Physics professor Roger Toomey awakens to find himself lofted to the ceiling, then rotates without moving and drifts down to bed beside June, his sleeping wife. She is aghast when he shows her this ability, and more so when he tucks her under his arm and flies around the room. Thereafter, he sleeps on the couch, held down by ropes. The department secretary gets the jitters and resigns when he accidentally floats out of his seat in a crouching posture. He writes to several notable experts who either ignore his report or send him a sniffy dismissal. In a mildly amusing scene he trips on stairs at the university:

> his autonomic system took over, and leaning forward, spread-eagled, fingers wide, legs half-buckled, he sailed down the flight glider-like. He might have been on wires.
> He was too dazed to right himself, too paralyzed with horror to do anything. Within two feet of the window at the bottom of the flight, he came to an automatic halt and hovered [p. 303].

Students witness this odd bird and rumors begin to fly. Soon he is close to being thrown out on his ear. On advice from a physician, he tries out as a stage magicians, but after a rousingly successful week he is fired. Customers will grow frightened if they understand that he is legitimate, sure that he is using demonic powers. June has had enough, and leaves him. Roger kills himself. A plainclothesman finds his cold body floating inches above the bed.

> Wildly, his flesh crawling, he put his hand on the dead chest and pressed downward.
> Something snapped. There was a clean, thin crack, tiny but distinct, and the body dropped [p. 329].

It is an interesting moment, for at last the empirical method has been brought to bear—too late, alas.

In the revised version, Roger torments the senior scientist who returned his letter via his superior, precipitating his firing. At a conference where the celebrated Linus Deering holds court, Roger levitates at the back of the audience, driving his foe to fury. His method succeeds when Deering, with the help of the FBI, offers him a post as Associate Professor of Physics, funded for full-time research on levitation. By showing the effect to Deering, and then denying that he had, Roger was able to provoke a defense in the skeptic against the man's only other explanation—that he was insane.

In principle this is an approach parapsychologists could use in the real world, but alas psi is too skittish and unreliable, however strong the accumulated statistical evidence is. What's more, Asimov makes a remarkable admission in his comments on the two versions of the story: "If you came to me ... and demonstrated that you could levitate ... I would probably proceed to disbelieve my eyes. Sorry" [p. 347].

An astonishing admission from a rational scientist, however typical; it is another benefit of psience fiction that we have this observation on record.

Algis Budrys, *"Riya's Foundling" (1953)*

At 22, Budrys (1931–2008) was a bright young writer patently intended for great things. He did produce some very fine work. One was *Rogue Moon* ([1960]; his own superior title was *The Death Machine*, about a teleportation device between Earth and Moon that created a double at the far end to investigate a dreadful alien artifact that left the visitor, and his telepathically linked original, in almost certain risk of insanity). Another was *Michaelmas* (1977), a splendidly solid novel about a benign newscaster who rules the world with secret aid of his AI, until he meets a mentality from the Multiverse. Budrys ended his life sadly as a stooge for the publishing wing of the Scientology cult, his creativity long lost to him.

In this very early jape[2] he presented a post World War Three boy, Phildee, with the ability to twist between realities via a Reimann warp. (Budrys surely meant Riemann, not Reimann, throughout, for the great German mathematician Bernhard Riemann whose geometric work on complex analysis is probably what he had in mind.) "Phildee" might have been a joking nod to Philip K. Dick, whose fiction took just such turns. The boy is a clever telepath.

> Abruptly, the Reimann fold became a concrete visualization. As though printed clearly in and around the air, which was simultaneously both around him and not around him, which existed/not existed in spacetime, he saw the sideslip diagram.
> He twisted.

And finds himself in a world of gene-engineered bison, who also communicate by a kind of telepathy he finds he can share. Riya Sair is an aging female of no interest any longer to males. She yearns for a male who will sire her a calf. Seeing Phildee, she concludes, as it were on genomic autopilot, that he is a suitable foundling.

Phildee's probe swept past the laboring mind directly into her telepathic, instinctual centers....

Soft as tender fingers, gentle as the human hand that smooths the awry hair back from the young forehead, Riya's mental caress enfolded Phildee.

She nurtures him with loving licks, tries to teach him to eat grass, looks forward proudly to the thousand mile walk to the southern range. Phildee knows this is impossible for him. "He adjusted his reality concept to Reimannian topology. Not actually, but subjectively, he felt himself beginning to slip Earthward." And arrives home with his new mother.

"Riya nuzzled her foundling. She looked about her at the War Orphans' Relocation Farm with her happy, happy eyes."

It is a sweet, primitive parable, moving forward into a drone of endless human war and death, psience fiction only by the skin of its buffalo teeth.

Cordwainer Smith, "The Game of Rat and Dragon" (1955)

In the 1950s and early 1960s, the most marvelous and weirdly poetic short stories and even novellas with a psi aspect were those by Dr. Paul Myron Anthony Linebarger (1913–66), who wrote as Felix C. Forrest, Carmichael Smith, and most evocatively Cordwainer Smith. He was fluent in Japanese, Chinese and some other languages, and his fiction often worked by parallel with the great literatures of those nations. A colonel, and expert in psychological warfare, he understood more about the realities of twentieth century guerrilla war than many of the generals who botched America's dealings with Korea, Vietnam and China. Some of this expertise works its threads through his short stories, as does his sympathy for the formerly enslaved people of color who are figured as the Underpeople, gene-engineered from animal stocks—a narrative element that might not please some of his readers, then or now. Both the mighty Lords and Ladies of the immense interstellar Instrumentality, and many of the insurgent Underpeople, are telepathic and have other psi gifts.

Perhaps his most famous early story was "The Game of Rat and Dragon," although later tales such as "The Dead Lady of Clown Town," "The Ballad of Lost C'Mell," and "A Planet Named Shayol" are richer and more musical. But the story of the warrior pinlighter Underhill and his marvelous, murderous Partner the Lady May in their endless telepathic battle against the terrible Dragons, seen by May and her kin as cosmic rats, is simple, pure and memorable. Starships planoforming between worlds in a kind of lightless hyperspace are open to ferocious attack by these denizens. Only the minds of humans can find the enemy and the instant reflexes of the Partners, linked by pin-sets, psionic devices, can act swiftly enough to win an encounter, blasting the foe in that darkness with light fiercer than the intensity of suns. Only slowly do we grasp the faintly shocking nature of the erotic bond between humans and Partners, who are cats: "A few hundred miles outside the ship, the Lady May thought back at him, "O warm, O generous, O gigantic man! O brave, O friendly, O tender and huge Partner! O wonderful with you, with you so good, good, good, warm, warm, now to fight, now to go, good with you" [p. 172].

Later, after the successful sortie, Underhill asks anxiously after the Lady May. A nurse scowls at him. He tries to probe her, but she has no skill with telepathy.

> Suddenly she swung around on him.
> "You pinlighters! You and your damn cats!"
> ...He saw himself as a radiant hero, clad in his smooth suede uniform, the pin-set crown shining like ancient royal jewels around his head...
> "She *is* a cat," he thought. "That's all she is—a *cat!*"
> But that was not how his mind saw her—quick beyond all dreams of speed, sharp, clever, unbelievably graceful, beautiful, wordless, and understanding" [p. 175].

Yes, this is a transcendent image of human and cat in utmost mental harmony. But it is also disturbing, in a way that Cordwainer Smith was candid enough to confess: "Where would he ever find a woman to compare to her?" A woman, that is, beautiful, wordless, and understanding. Psience fiction as a song of sexism.

On the other hand, there is no reason why female pin-lighters like the little girl named West (p. 168) would not feel exactly the same way about their magnificent tomcat Partners. Hers, after all, is Captain Wow.

Brian W. Aldiss, *"Psyclops" (1956)*

In a curious reversal of MacLean's "Defense Mechanism," this early Aldiss story follows the telepathically guided very premature awakening of a six-month fetus in the womb. A twin sister is all but dormant. Stirred to consciousness by his father, who is trapped by blue aliens, and with a gangrenous leg, on the planet of a star many light years distant and seeking psionic help the only desperate way available, this all but disembodied mind in darkness works his way from solipsism to a sense of I and Thou, with many slips back into disbelief. The tale is a small *tour de force*, like "The Mindworm" really unparaphrasable because so much of the message is the unorthodox format of the medium. It is also, the reader sees, a kind of grim Woody Allen jest, with the premature baby aware of dangers to the ship he is traveling on, to his mother, to his sister, even if his father is doomed and finally out of reach.

His grasp of the complexities of the world is rudimentary, and so is the *Aliene*sque solution he finds:

> Maybe I could call more easily if I was outside, in the real universe. If I turn again.
> Now if I *kick*...
> Good. Something yielding.
> Kick [p. 41].

J. T. McIntosh, "*Empath*" (1956)

Bringing a slightly different emphasis to a psience fictional trope—empathy of shared feelings rather than the usual telepathic communication of ideas—J. T. McIntosh's novelette received an Honorable Mention in Judith Merril's annual *Year's Greatest Science Fiction and Fantasy* anthology.

When wealthy but frail empath Robert Green falls off the second highest roof in a nuclear-bomb-threatened city, beautiful young Betty Lincoln (*virgo intacta*, as it turns out) is framed for pushing him to his death. What's an empath? Tim Green, Bob's younger brother, aide to the police, and also an empath, explains: "It isn't telepathy. No one can pick words or ideas out of another person's brain—yet. It's a sensitivity to atmosphere, to aura. To feelings.... You can't tell what's going to happen, or what people are thinking. You can just feel things—and guess" (p. 28).

Tim realizes Betty's innocence of the crime at once. Set free, she returns to her apartment and finds the man who flung Robert to his doom. He has a gun, so Betty outwits and outfights him. Could it be that she, too, is an empath? Why, yes.

A group known as the Circle, empaths and otherwise, are eager to kill the few on the side of law and order. In a display of non-telepathic and non-clairvoyant psi, although the logic of the story suggests that really what is involved must be at least partly clairvoyant, the remaining Circle empaths and Tim hunt each other from one part of the underground city to another. Tim finds a place to hide from them "because I had to." How empathy manages to achieve any of this is not explained: "It wasn't a battle of minds. It was a conflict of emotions, almost tangible emotions, feelings which could be passed from one mind to another" (p. 37).

Nor is Betty's ability to direct the police by jabbing at points on a map explained—a somewhat well-known method of scrying, but requiring rather different psi skills. At length the bad guys are shot dead (no tedious reading of rights, arrest, judicial hearings for these malefactors, and no sign of regret about this either), and Betty is united to Tim, whom she loves ardently after only a few minutes in his company. Such is the power of empathy.

In short, McIntosh seems to be making this up as he goes, without quite the same dream-logic impetus van Vogt provided in his early fiction. Why was Robert Green trying to gain Betty's attention before the Circle thugs took them both up to the room? Presumably he detected her nascent empathy and hoped to engage her help, but that does not appear consistent with what the Green brothers know of empathy. But only one person in half a million has this gift, and most become psychotic. Empathy of this kind has only been known for five years. Perhaps it is a side effect of the atom bombing of several South American cities, giving rise to the fashion for rich "moles" to live underground and poor "angels" to be exiled to the top of skyscrapers where any nuclear weapon effects will be most brutal. Tim explains the group dynamics of the thing: "[W]e empaths are a different kind of people, even if we don't learn this until comparatively late in life. What do people do when they discover they're different, in some way superior to ordinary people? They try to take over the ordinary people" (p. 29).

Only the few, apparently, are able to resist this fascistic temptation. Luckily Betty is one of them, and so was Bob and so is Tim. Also luckily, she has fallen in love with Tim, and he is bound to do the same in return, once the feeling between them "bounces back and forth between us and keeps getting bigger" (p. 44). The value of this story, then, is as an example of how not to design psience fiction, even when you have what seems a rather nice new variety of psi power to put into play.

Poul Anderson, *"Journeys End"* (1957)

Here, as the finale of our voyage through some characteristic psi tales of the 1950s (and one from the very end of the 1940s), is a stinging tale from that polymathic worker of sf miracles, Poul Anderson (1926–2001). He explores the plight of the only telepathic man in the world, as far as the poor fellow knows—until, on a swiftly moving train, he catches the reciprocal female thought from a train speeding in the other direction. What to do? Telepathic range for Norman Kane is only a few hundred yards (although this is far from the case in actual scientific studies, which observe psi effects from one side of the world to the other). He has been alone in his gift for fifteen years; he tries personals in the major newspapers (today, of course, he would have Facebook and Google and the possibility of finding Her in an hour or two).

Anderson follows Kane in his cigarette-puffing journey to its bitter end—as signaled in the title, which you'll notice has no apostrophe before the 's'.... He is in love with this woman of his dreams, and inundated as ever by the stray and deeper flux of thoughts and anxieties and often horrible plans and responses of those he moves amongst. Like this:

> —*doctor bill and twinges in chest but must be all right maybe indigestion* (p. 128).

But worse yet, the terrifying vertigo of:

> —*flayed and burningburningburning moldering rotted flesh & the bones the white hard clean bones coming out gwtjklfmx—*

Norman, snatching at recovery, draws on his memories of advisor Father Schliemann, like "a deep well under sun-speckled trees, its surface brightened with a few gold-colored autumn leaves ... a sharp mineral tang, a smell of the living earth" (p. 130), an advance glimpse, as we shall see later, of the farmer mystic encountered by young telepath David Selig in Silverberg's *Dying Inside*. Kane has been tormented until he ran away from home by his insanely doctrinaire and sadist anti-papist father. There really are a lot of problem fathers in psience fiction.

He moves toward the university library where his search will continue, and stumbles upon the young woman, twenty, not beautiful but not unattractive, kind, opening to his mind as his opens to her. They are horrified.

> —*buried uncleanness & in the top of my head I know it doesn't mean anything but down underneath is all which was drilled into me when I was a baby & I will not admit to ANYONE else that such things exist in ME*—[p. 136].

They cannot tolerate each other's presence.

Perhaps this is the true reason why psi is so evanescent, cloudy, vivid in moments of extreme need but then fading into air.

24.
Mark Phillips (Randall Garrett and Laurence M. Janifer), *Brain Twister, Impossibles* and *Supermind* (1959–1961)

John W. Campbell's enthusiasm for psience fiction, then, reached what was either a pinnacle or a *reductio ad absurdum* between 1959 and 1961, the years when *Astounding Science Fiction* was rebranded as *Analog Science Fact-Science Fiction*. Three light, comic police procedurals of a psionic bent were serialized under the pseudonym "Mark Phillips," and were collaborations between an *Astounding* regular, Randall Garrett (1927–1987), and Laurence M. Janifer (1933–2002).

Brain Twister, Impossibles and Supermind (1959-1961)

A serial version of *Brain Twister* (1962) appeared in *Astounding* under the title "That Sweet Little Old Lady" in September and October 1959. Its immediate sequel, "Out Like a Light," subsequently published as the book *The Impossibles* (1963), was in *Analog*, April-June 1960. The concluding serial, "Occasion for Disaster" (*Analog*, November-December 1960 and January-February 1961), came out as *Supermind* (1963). The first dealt with telepathy, the second with teleportation, and the third with a sort of mental static afflicting the world rather as computer viruses, spammers and scammers would mess with the Internet thirty or forty years later.

These frothy romps were generally liked by readers, if Campbell's "Analytical Laboratory" calculations are to be trusted (and perhaps they should not be), but were tailing off by the third. Were *Analog* supporters losing interest in psience fiction, or were they growing bored by the frivolity of the Mark Phillips approach? Certainly the joint authors' approach to storytelling was surprisingly padded. To take a random example:

> "Can Sir Lewis get me all the data on that tonight?"
> "Tonight?" Luba said. "It's pretty late and what with sending them from New York to Nevada—"
> "Don't worry about that," Malone said...
> "He'd have to send the originals," Luna said.
> "I'll guarantee their safety," Malone said. "But I need the data right now."
> Luba hesitated.
> "Tell him to bill the FBI," Malone said. "Call him collect and he can bill the phone call, too."
> "All right, Ken," Luba said at last. "I'll try."
> She went off to make the call, and came back in a few minutes.
> "O.K.?" Malone said.

And on and on in the same increasingly irritating fashion. Sir Lewis, by the way, is Sir Lewis Carter, President of the American Society for Psychical Research—a tip of the hat to James Blish's vastly more interesting psience fiction novel *Jack of Eagles*, where Sir Lewis Carter was the evil astronomer and head of the Psychic Research Society. In the third volume, the Major of New York is Amalfi, also by coincidence the name of the ancient immortal Major of the levitating city in Blish's *Cities in Flight*. Garrett could be brilliantly funny—his spoof of *The Demolished Man* approaches genius, ending "Craye D'Courtney and Barbara, respectively, are The (ter of Reich"—but not so much in this sequence. On the other hand, the trilogy might be regarded as an early version of psi as hilarity in the mode of the genuinely comic *Ghostbusters* movie (1984 version). It is not impossible, or a brain twister, to imagine it forming the basis of a successful television or Netflix mini-series.

FBI agent Kenneth Malone is assigned a new task. His Bureau Chief,

24. Mark Phillips (Randall Garrett and Laurence M. Janifer)

Anthony J. Burris, tells him: "They could be anywhere.... They could be all around us, Heaven only knows.... We're helpless.... There's a spy at work in the Nevada plant, Kenneth. And the spy is a telepath" ("Little Old Lady," Part 1, pp. 10–12).

One other telepath has been identified, but he was psychotic and is now dead. At the Project Isle plant, in Yucca Flats, Nevada, the first non-fueled space drive (another of Campbell's *idées fixes* at the time) is under development. A Westinghouse psionic device (Campbell's major *idée fixe*) is capable of detecting whether someone's brain has been spied upon, but it is so far too crude to identify the spy. For that, another telepath is needed.

The mild ingenuity of the ruling conceit in the sequence is that the first powerful telepath is discovered by an FBI agent in a lunatic asylum, where she is indeed a sweet little psychotic in her sixties, Rose Thompson, who maintains that she is a telepath. Not only that, but she's convinced that she is Queen Elizabeth the First, still alive after hundreds of years and eager for her loyal colonial subjects to do her bidding. The agent and his new FBI associate Thomas Boyd learn quickly to fall in with this delusion, sometimes wearing the full regalia of the Elizabethan period in blazing American summer, and gaining knighthoods for their service to royalty. Is it mere accident that Boyd in that regalia bears a striking resemblance to Henry VIII?

After Keystone Kops escapades, Her Majesty leads them to the psychic spy, Willy Logan (a reference, perhaps, to playwright Arthur Miller's depressive salesman Willy Loman?). Young Logan is a mental patient in Las Vegas. He has been tortured by his psychiatrist, Dr. Frederic Dowson, who has used Willy's almost unique skills to filch information on the space drive and sell it treacherously to the Russians. This is a somewhat amusing version of the standard spy plot of the end of the 1950s, and the element that makes this novel of interest as psience fiction is Her Majesty's explanation for Willy's apparent catatonia. It is a benign version of the common parapsychological account of links between poltergeist kinetic effects and troubled adolescence:

> "In the first place," the Queen said patiently, "Willy isn't catatonic. He's just *busy*, that's all. He's only a boy, and ... well, he doesn't much like being who he is. So he visits other people's minds, and that way he becomes *them* for a while.... He didn't want to be a spy, really," she said, "but he's just a boy, and it must have sounded rather exciting" [Part 2, p. 137].

Malone, meanwhile, attributes his success in this and earlier difficult cases to being "lucky." Sometimes, though, lucky is just another word for psi.

Ψ

The other serials follow this general pattern. "Out like a Light" references the uncanny experience of watching someone teleport, simply vanishing like a light switched off. Locked red 1972 Cadillacs are being stolen. Malone finds one,

Brain Twister, Impossibles and Supermind (1959–1961)   111

is knocked unconscious, it is gone when he wakes—but he finds a notebook dropped by the culprit, filled with odd and cryptic notations. The names of members of an Italian-American teenage gang, the Silent Spooks, are listed with one exception. This is their leader, Miguel (or Mike) Fueyo. Already in custody, Mike is questioned by Malone and mockingly vanishes from the interrogation room. So: a new psi power is in town, teleportation. With such a gift, many robberies could be pulled without leaving a trace. This pattern is confirmed.

Malone's luck pays off again, it seems, when he meets beautiful Dorothy Francis at the precinct station and arranges a date with her. He gets roguishly drunk, wakes the next day without the notebook. Of course Dorothy is really Dorothea F. Fueyo, Mick's sister. Her Majesty calls, explains that she has been following the whole case via rapport with Dorothea via Ken Malone. Dr. O'Connor, the supercilious space drive expert, jots down the equation for PK, again, it seems, in mimicry of Blish's Blackett equation. K is an unknown constant, which equals mass m in grams by d, distance in centimeters, divided by force f in dynes by t^2, time in seconds squared. "What we are measuring here," he explains, by f, "is the strength of the subject's personality ... another factor, t, is the span of attention of the individual—the ability of the subject's mind to concentrate on a given thing for a span of time" ["Light," Part 3, p. 132].

There is also, naturally, *volume* of attention, the capacity to conceive and hold awareness of the space around a teleport. Handcuff Mike or one of his confederates to a radiator in the precinct station and he would not be able to skip out. But how did a street gang get hold of this secret research information? It seems Mike had the gift, and worked out how to notate it, using this compressed knowledge to teach his cohort the trick. Dorothea teaches Malone how to do it, and he's suddenly a teleport as well. And he wangles Mike an FBI job as a special-case Fed. Does the whole world change as this knowledge spreads wildly between nations, like the trick of jaunting did in Bester's *The Stars My Destination*? No—there's still another book to go.

Ψ

"Occasion for Disaster," aka *Supermind*, ups the consequences of psi on the loose. Malone has become prescient—precognitive—and everyone agrees that the world is going to hell. Why? Here the witticism is three spies accused of causing the mental static by putting psychoactive drugs in the water: Brubitsch, Borbitsch and Garbitsch. Malone contacts Maigret of the Paris Sureté and Poirot in Belgium, although there is no mention of Clark Kent in Metropolis. Something along the line of ghosts are haunting the Big Apple. Could it be coordinated telepathy reaching non-telepaths throughout New York? Det. O'Connor doubts it:

> Very few [reported cases] have been written up with any accuracy, and those seem to be confined to close relatives or loved ones of the person projecting the

message ... the projector of such a message is in dire peril ... the amount of psionic energy requires for such a feat is tremendous. Usually, it is the final burst of all the remaining psionic force immediately before death ["Occasion," Part 4, p. 136].

Malone contacts the Psychical Research Society. He has previously had a hot time with their secretary, Luba Ardanko; Lou has gone too, breaking their arrangement. In a burst of insight, Ken Malone understands that the PRS membership is telepathic. The organization has been investigating the paranormal since 1880, and they have learned to use and control psi in all its forms. "[A]ll of you out there," he tells them mentally, driving toward their mansion headquarters, "are responsible for what's happening to this country and all of Europe and Asia—and, for all I know, the suburbs of Hell" (p. 154). Their psychic interventions have turned the world into a bloody shambles. Yes, some bad people have been ruined, but "You've done more damage in two weeks than those fumblebrains have been able to do in several myriads of lifetimes" (p. 156).

He is met by Andrew J, Burris, Director of the FBI, and Sir Lewis, and Luba. He's got it all wrong, of course. The psiontists had to stop the motor of the world, or at least gum up the works for a while, because otherwise some fool "was going to start the Last War" (p. 161). There's an explanation, but not a very persuasive one, and Malone doesn't really understand it, understandably. And he is given deep and total access to Luba's mind, because she knows he is in love with her and it just wouldn't work. Malone is convinced:

> When a man knows and believes that someone actually *is* superior—then, he doesn't mind at all. He can depend on that superiority to help him. And love, ordinary man-and-woman love, just can't exist.
> Nor, Malone told himself, would anyone want it to. It would, after all, be too damned uncomfortable [p. 165].

Thus, psi gender politics in the United States at the opening of the Sixties. It was not entirely different in Britain at that time, as we shall now see.

25.
Arthur Sellings, *Telepath* (1962)

"Boy" telepath meets "girl" telepath at a London party in the future year of 1975 and neither knows that's what they are, but fireworks and Catherine wheels ignite their hearts, bearing them off to a quiet region where the noise of the party dropped away to a sound that was no more than the soft breaking of surf.

Here they were alone, beyond time and space, trembling from the shock, but oddly calm. Calm and naked. Stripped of words and misinterpretations ... they shared the same universe, one which until now had not existed.... It was like waking into the world after a long night of dreams [p. 11].

For a few minutes. Then the female telepath, Claire Bergen, an artist who has never shown her work, runs away into the night. The male telepath, Arnold Ash, who was a poet before he decided it was all futile and became an advertising copywriter, tracks her down, and then has to do it all over again in her country retreat. More psychic explosions and defensive hostility ensue. An elegant and slightly surprising romantic comedy opening for a 1962 Ballantine Books psience fiction novel, in flavor a little akin to John Wyndham (whose *Trouble With Lichen* came out from Ballantine in 1960) or Brian Aldiss (his *The Primal Urge* did the same in 1961) or John Brunner (*The Whole Man*, same publisher, 1964). It is intriguing that these very British novels were first released by an American press.

What's particularly interesting about Sellings' approach to telepathy is his focus on gender conflict from what amounts to a fifties' standpoint. The clichéd binary contrast proposed in that period as the basic grammar of both gender and sex is carefully undercut even as it remains valorized. For a start, Sellings' presentation of psi communication shows a merging of minds, an enlarging. Vic, a rather oafish "blokey"[1] friend of Ash's, is the first person to whom he chooses to reveal his new ability, so he gets a lot of standard blustering about this woman giving him trouble, so to hell with her. Oddly, Vic in reassuring mode refers to himself as "Auntie Vic, dear."

In their often caustic exchanges, Arnold and Claire are posed as polar contraries to the typical male/female constellations: he disapproves of the analytic, which she defends and exemplifies. Yet after a burst of psychic oneness that turns into passionate sex, Claire is "prim" and calls off their decision to work together on extending this gift to humankind. Arnold is sarcastic: "You're just playing the outraged female" (p. 65). This is true, as she agrees, but the terrified Zeitgeist of the males with whom Arnold is surrounded is candidly acknowledged when they discuss joint accommodation. His "fiercely bachelor block" (not gay, one gathers) is sketched with sharp mockery:

> A good half of the occupants were divorced or separate. All of them lived in a kind of group fantasy. They saw themselves as victimized by womanhood. They had all *suffered*, even the not yet married whom the ex-marrieds drew into their fantasy. A woman on the premises was overtly tolerated—in actual fact she was welcomed as a kind of ritual sacrifice ... [yet] a woman in permanent residence would have been a symbol of defeat [p. 67].

They are both traumatized when an 11-year-old voiceless girl, Sally, taught telepathy by Claire, is raped and beaten (surely, for the narrative, a kind of ritual sacrifice) by a manipulative brute who kills himself before he is apprehended. Nothing is simple or easy in this exploration of the consequences of emergent psi, even if, as the novel's McCarthy-hounded Jewish sociologist Michael Green claims, psi is a "countermutation" evolved to deal with mass society's imposed uniformity and psychiatric hazards. Yet Arnold's attempts to reach the populace with word of their breakthrough are comically inept. And if rape or even oafish Vic's newly emergent psychic sexual predation are all-too-common consequences of fearless empathy in the victims, what of Arnold's enforced treatment of a space scientist driven insane by helping British military damage or kill "dozens? ... hundreds? ... of space-mad kids" in a lunar barracks (p. 125). (By 1975, the United States, Russia, China and Britain have colonies on the moon, people use microwave ovens to make coffee but call them "high-frequency cooking units," and watch "stereos" in three dimensions but not by using holograms.)

The last third or so of the book has Arnold in Woomera, Australia, foraging inside the mad scientist's brain. Finally he has the man returned to working order, and is sent back to London, where news media already know he is of interest but not much more than that. Neither do we, really. Then he is kidnapped by pleasant James Bond villains who set him free after an Israeli spacecraft lands on the Moon without needing boring Newtonian reaction forces. It seems rather like painting by numbers at this stage, although there is a constant awareness that Arnold and Claire have discovered something of epoch-changing significance, and its development is just on sabbatical for the moment.

To some degree, Sellings' picture of a military-political reaction to proven cognitive psi capacities resembles Wilson Tucker's in *Wild Talent*, but it is far more melodramatic than the reality of the U.S. government ESP program. Reports by remote viewers and their advisors—Russell Targ, Joe McMoneagle, Paul Smith, and others—reveal a very much more routine and boring procedure.[2] Still, *Telepath* has the virtue of being concerned with ethics in the military-industrial-intelligence state that was already well established in the early 1960s, and far more pervasive and intrusive today. The question the novel leaves us with is this: Must command of psi enhance the power of the state, criminals, and grasping corporations, or might it strengthen community and the status of the individual?

The novels end with the return of Professor Michael Green (née Moishe Grunberg) to expound a link between the reactionless starship (humankind will now explode into the galaxy) and psi communication (only this can hold together the ancestral species dispersed on an interstellar scale). Then Claire and young Sally return after a refreshing time with the salt of the earth in

Ireland, "looking after the animals and cutting turf ... for fuel," knitting a sweater for "Arn," "sheared, spun, woven, dyed with local dye," and carrying a batch of freshly painted canvases in a more involved, less analytic mode (pp. 158–59). This being the very early '60s pretending to be 1975 and having no oral contraceptives and apparently no condoms, an additional presence waits to share their joint life. And this actually being barely out of the 1950s, there's "something else we'll have to see to," Arnold points out. "And quickly" (p. 160). Wedding bells ring beyond the closing page.

26. Keith Woodcott, a.k.a. John Brunner, *Crack of Doom/The Psionic Menace* (1962–1963)

In 1962, when John Brunner published this short novel in *New Worlds* under the pseudonym above, editor John Carnell introduced it thus: "A reader once stated that the right type of psi story had never been written for British science fiction. We contend that Keith Woodcott's two-part serial is just right internationally and introduces a slan-like atmosphere into what has become a highly controversial theme."

The following year the novel was reprinted as a lowly Ace Double, still under the pen-name, but with the title changed to *The Psionic Menace*. It was backed by *Captives of the Flame*, the remarkable opening volume of Samuel R. Delany's trilogy *The Fall of the Towers*, where it exchanged the vulgar Ace title for *Out of the Dead City*.) It is not clear why Brunner used a false name for *Crack of Doom* and several other short novels, but it surely was not because he was afraid of controversial themes, which became one of his hallmarks.

Brunner deals here in one of the obvious possibilities were telepathy ever to become consciously available, but with no stop switch. Mind readers would perhaps be inundated by a nonstop screeching and babbling, rather like sitting in a cinema or train while everyone else nattered away loudly on their cell phones. The psions deal with this problem to some extent by relocating into rural communities, as far from the madding crowd as they can afford. Presumably it is easier to modulate reception from other telepaths, who at least are aware of the problem. What's more, some typical hazards of *Homo fallax* (the Deceptive Human) are remediated by the psychics' built-in inability to tell consequential lies, because you give yourself away even before you open your mouth—and most psions don't bother with oral speech in any case.

The only kind of human able to circumvent this propensity for self incrimination in the company of a psion is the *psinul* variant, lacking even the residual psi available unconsciously to most humans, a deficit that crops up only once in 16 million births. (Psions are more frequent, but only just, about one per two million.) Philip Gascon only learns that he is a psinul when he finds a psion child running screaming in the night, terrified by a kind of roaring telepathic shout announcing that something dreadful is coming, a monstrous and overwhelming catastrophe. At 25, Gascon is studying for his doctorate in cosmoarcheology, on the trail of the Old Race that once ruled more than forty worlds but vanished 100,000 years earlier, leaving only enigmatic structures and tools so far all but useless to the human species.

This provides a handy pretext when the authorities of a kinder, gentler Earth of the future send Gascon to Regnier. This world falls inside the domain of the human Starfolk, who hate psions and regard these mutant variants their particular rivals as evolution's flag bearers: "Earth—old, wise, tolerant—tolerated psions. The Starfolk, seeing them as rival claimants, could not; they encouraged psion persecution and extended their mistrust to Earth and its associated planets where psions were permitted to exist openly" [Part 1, p. 25]. The Starfolk traders, derived from human stock but living in space, sustain radiation-damage, particularly sterility, so they bargain for the regular transfer to their star cities of a selection of fertile youngsters from Regnier and other planets.

The vulnerability of telepaths to a sort of psychic overload, reinforced by positive feedback from their own terror and incomprehension, is a topic worth exploring in psience fiction. Must psions be driven almost insane by such an effect, ending with hands uselessly jammed over ears and mouths open in pain, like the figure in Edvard Munch's expressionist painting *The Scream*? Or would they adapt, gradually lose their psychic hearing, like workers in intensely noisy factories, or artillery officers and crew?

Brunner shows them struggling to adapt to this shout from an unknown source, this affliction that seems something like a really bad toothache that gets more and more painful but can't be treated. Gascon's inability to register any psi signal allows him to take the lead both in studying this unprecedented mental bombardment and its political consequences between the starborne Starfolk and both the psions and whipped-up psion-haters of Regnier. The book was written less than two decades after the end of the Second World War; parallels with the fate of Jews and other victims of the Nazi Holocaust are apparent. (Ace's retitling it to *The Psionic Menace* was typically tone-deaf in that epoch; would they have published a novel titled *The Jewish Menace*?) Brunner was only ten years old when the murder of those millions came to an end but his detestation of political scapegoating and the threats of racism and nuclear devastation in the years that followed is declared plainly in his fiction.

Earth's Executive, named Fold, offers Gascon possible explanations for the psi "shout":

> It may be the signal of agony enduring in a psion pogrom on Regnier—the last we heard, such a pogrom was a strong possibility. Alternatively, it might be a signal from the future; there are theoretical grounds for believing psionic communication need not take place in present time. In that case it may *really* be a warning of the end of everything...
>
> But the most acceptable hypothesis is that it's from the past, and that the Old Race left a psionic warning of their fate for those who might follow them [Part 1, p. 31].

The development of a prickly, space-centered Starfolk way of life is a routine example of cultural fragmentation and specialization. But what caused the slow emergence, over half a millennium, of the psions? This is seen by Fold as a "response to an environmental requirement.... What acted on our minds as incident sunlight acted on the primal eye [in prompting the evolution of vision]? I don't know" (39). A plausible explanation is offered at the end of the novel, after the nature of the frightful "shout" is uncovered. The Old Race still exists, in a sort of immaterial form, and all return to their own original star to die—each taking its subjective universe into oblivion with it. It notifies the galaxy of its impending demise by a psionic call, amplified in this case by human psions. Fold notes that "Perhaps it was under the impact of this signal that we began to select for psionic awareness—after all, it makes sound biological sense to respond to a subconscious warning of disaster" (Part 2, p. 120). Arguably, given its atrocious consequences for the psychic humans, it would make better evolutionary sense to reward selection for psi deafness. On the other hand, being psinul has dubious consequences, although perhaps it speaks to the larger-than-average Asbergerish fraction among sf's readership:

> A psinul ... has no power of psionic perception ... not even the rudimentary talent which millennia of genetic selection has encouraged among the communities of socially-organized man. Other people are in a sense mysterious to you; you can't instinctively make the judgments they make, and you're compelled to resort to intellectual analysis, which is far more difficult. You feel this lack—you compensate somewhat by leaning over backwards ... to display social responsibility [Part 1, p. 30].

Still in his twenties when he wrote this only slightly better than routine short adventure novel, Brunner tries for something involving and thoughtful but fails to reach anything much beyond the kind of fake transcendence that would mark many early Gene Roddenberry *Star Trek* episodes half a decade later. The Old Race lured by the call of their dying companion look like "veils, filmy, indistinct, of multiple soft colors, drifting in from the distant stars towards this planet of all planets, patterns of organized non-physical energy awareness beings" (Part 2, p. 117). Still, his treatment of the psi-gifted as a new persecuted minority, whose superior abilities actually reflect the better angels of our nature, is intriguing, an advance over that of van Vogt slans and even Wyndham's chrysalids. A more nuanced portrait of a wounded psychic was provided by Brunner two years later, and under his own name; we turn to that novel next.

27.
John Brunner,
Telepathist/The Whole Man (1964)

If *Crack of Doom* was just a pseudonymous workman-like exercise in genre psience fiction, *Telepathist* (published in the U.S. as *The Whole Man*, perhaps a superior title) was an act of redemption that produced a fairly convincing attempt at an adult novel. It began as the first of a projected template series with "City of the Tiger," a gaudy sword-and-sorcery novella with a startling ending. The Conan-lookalike viewpoint character unmasks himself in the closing stanza and is revealed as the brilliant, crippled, stunted psychologist Gerald Howson, Psi.D, curative telepathist first class.

His profession, dependent on his rare psi gift, treats people in the grip of incurable psychoses or similarly resistant mental deformations by entering with them into their dreams and leading them via imagined transactions in imagined worlds back to insight, sanity and health. (A related idea was explored in Roger Zelazny's "He Who Shapes" two years later, and by Greg Bear in *Queen of Angels* in 1990, although their guided hallucinations were machine-implemented rather than psionic.)

While "City of the Tiger" was effective and liked by readers, Brunner found he was not attracted to the task of reusing the template, however ingenious, again and again (as he had done in the time travel/alternative history trio collected as *Times Without Number*. Instead, he stripped away much of the derring-do of the originating tale[1] and went back to Gerald's miserable childhood and early adulthood, his deformed face "like an idiot child about to vomit!" (p. 61), following him through psi training and then into several of his most consequential cases. These are chosen not for their stand-alone value but as components in an intriguing and poignant biography.

His childhood is dire; his hastily departed father was a detested terrorist, his mother uninterested in this wretched little creature. He was stunted and twisted and his voice never broke; he did such small jobs as he could find and tolerated the endless gibes:

> At twenty, Gerald Howson was convinced that the world which had been uncaring when he was born was uncaring now, and he spent as much time as possible withdrawing from it into a private universe where there was nobody to stare at him, nobody to shout at him for clumsiness, nobody to resent his existence because his form blasphemed the shape of humanity [p. 24].

His blasphemous shape, in fact, is an echo of those many sf novels and stories where a mutant is physically deformed or just different and additionally, as a by-product of radiation damage or the like, a psychic superbeing. Howson's telepathic abilities emerge when he is twenty, as, as this novel, they often do in the psi gifted. No explanation is offered for this rare novelty, although the physiological basis of psi is known: hypertrophy of the organ of Funck, deep in the brain. Tragically for Howson, this neural complex is adjacent to a map of his bodyplan, and the excrescence has not only warped that architecture but absorbed much of its rightful territory. Even after Gerald's prodigious gift is discovered and he is taken to the Ulan Bator psi center and introduced to the proper medical use of his skill, he cannot be repaired without deleting both his psi power and possibly his sanity.

Eleven years later, a Psi.D. curative telepathist, he is an expert in helping free a catapathic grouping—a delusional state initiated by some brilliantly capable mind reader whose victims are trapped in a coma while their minds enjoy a fantasy akin to the immersion computer games that did not exist when Brunner wrote the book. Howson has achieved success in some thirty of these desperately difficult cases, in which he must insert himself into the collective narrative without being trapped. Now he is obliged to enter the Chinese City of the Tiger designed and hosted by another genius telepathist, diplomat Hugh Choong. Although he succeeds in the task, he understands belatedly that this was a deliberate engagement by Choong, and his self-confidence is badly damaged. He leaves Ulan Bator for a vacation return to his childhood territory, searching for his brutalized roots.

Brunner's representation of telepathy throughout is varied and almost plausible. Although he must present exchanges between telepathists in something like regular English syntax, he makes it clear that this is just for the sake of us non-psychic readers. In a clever narrative device, we are introduced to this device via Howson's discovery, as a youth on the run from police and criminals, of a young woman without hearing or speech:

> At first he could make no sense of the impressions he took from her mind, because she had never developed verbal thinking; she used kinaesthetic and visual data in huge intermingled blocks, like a sour porridge with stones in it.

While he struggled to achieve more than the first broad halting concepts of reassurance, she sat gazing at him and weeping silently, released from loneliness after intolerable years [p. 43].

Once again, here we find one of the key drivers in the process of psience fiction: minds isolated, separated from the rest of humankind, often despised or rejected, finding at last a kindred soul. And often enough messing up that opportunity as well, because humans are such difficult and complicated creatures. In the case of Mary, the deaf mute, Howson finds in her alexic mind a jumble of layers of memory and compensation:

> Half the girl's mind knew what her father was actually like: a dockland roustabout, always dirty, often drunk, with a filthy temper and a mouth that gaped terrifyingly, uttering *something* which she compared to an invisible vomit because she had never heard a single word uttered...
> But at the same time ... she maintained an idealized picture of him, blended out of the times when he had dressed smartly for weddings and parties, and the times when he had shown loving behavior towards her ... still further overlaid with traces of an immense fantasy from whose fringes Howson shied away reflexively, in the depths of which the girl was a foundling princess [p. 44].

This empathic sharing becomes, of course, the basis for Howson's career, and self-understanding, as a curative telepathist. In the climax of the book, almost ready to return to Ulan Bator, he shares another non-psionic person's abnormally gifted but rejected mental construct. This is an unprecedented multidimensional artform blending visual, musical and other components in ways that can only truly be appreciated by a brilliant telepathist such as Howson capable of entering the imagination of young suicidally unhappy artist Rudi— and, luckily, finally, of transmitting it to an ordinary willing audience.

This is presented to us in a jumble of fragments that are not necessarily effective for all readers, although it might work with synaesthetes who already experience colors as sounds and vice versa, with scents and thuds and every other sense medium added to the mix. One misfortune is Brunner's devising a futuristic imaging system—"wet fireworks"—that amounts to his *avant la lettre* invention of lava lamps, destined a decade later to be an item of parody: "a stream of opalescent bubbles began to work their way through the tanks in an irregular series of graceful loops..." (p. 181). The complete performance is perhaps like a three dimensional Disney *Fantasia* rendering of the score of Alan Hovhaness's "Mysterious Mountain," with clashing garbage bin lids and perhaps the stomach-testing perfume of haggis:

> Mountains gray in the tank, distorted as if looked at from below, purple-blue and overpowering; mists gathered at their peaks, and an avalanche thundering into a valley surrounded by white sprays of snow, as a distant and melancholy horn theme dissolved in Rudi's mind into a cataclysm of orchestral sounds and a hundred un-musical noises [p. 184].

For Howson, helping to bring this hidden sound vision into the grasp of an audience from whom it has always been hidden is satisfaction, even joy, enough. He has, Brunner assures us, "become a whole man."

28.
Dan Morgan,
The Sixth Perception Series
(1967–1975)

Between 1967 and 1975, British professional guitarist Dan Morgan (1925–2011) produced a quite solid psience fiction sequence: *The New Minds* (1967), *The Several Minds* (1969), *Mind Trap* (1970) and *The Country of the Mind* (1975). This is not to say that what it describes is an accurate account of psi events as they would be reported by scientists. For example, the opening of *The New Minds* takes us inside the mind of young Toby, as he lies somewhat painfully on something akin to an fMRI table, awaiting the appointed time for him to link telepathically to his twin Sid, a hundred miles distance in London: "the last time when we did It over twenty miles, we was both in bed for two weeks after, weak as kittens. They had to feed me with slops and that through a tube for the first three days, and it was too much trouble even to think" [Book 1, p. 9].

In reality, there is no aversive consequence from trying telepathic contact even between Earth and part way to the Moon, as astronaut Dr. Edgar Mitchell showed in Apollo 14. And remote viewers routinely "visit" locations as far away as the other side of the planet with no ill effects. But the point about psience fiction is not to stay within the current boundaries of known parascience (or we would be without *The Stars My Destination* and its routine teleportational jaunting, not to mention jumping through time with your hair on fire). It is not called *fiction* by accident.

As the twins lie curled in catatonia, their Portfield House sponsor Dr. Richard Havenlake, his colleague Dr. Rebecca Shofield, and psi-gifted associate Peter Moray struggle to save them, and save as well Barbara Graham, a young suicidal telepath on the edge of madness in an institution where she is about to be subjected to ECT, which will unintentionally burn out her telepathic abilities. Although the Havenlake team have managed to persuade the highest British authorities of their legitimacy, they are hampered by an arrogant upper caste pest, Edmund Powell, appointed to be their watchdog, and are dismissed as crazy by the psychiatrist in charge of Barbara—until they return to his hospital with official backing to free her, and find that nobody acknowledges her existence. Another, stronger, unknown psi master is abroad, manipulating minds and presumably responsible for the twins' collapse. Meanwhile, Havenlake's once delightful wife is deteriorating into a drunken, shrieking, malevolent harpy.

So far we have the basis for a dark and even thoughtful movie by, say, Hitchcock, akin to Byron Haskin's *The Power*. As it develops through four volumes, Morgan's psience fiction sequence proves to be more complex than that. When Barbara wakes in the custody of a sour, chunky woman with garish makeup, she learns who is responsible for her captivity. Entering a small odorous room, she finds a childlike very deformed person somewhat like a seal, 25 years old to her 21, with tiny perfect fingers but no shoulders, and literally adorable. "Victor, who in his exquisite helplessness needed her more than anyone in the world, who had brought her here to be his friend, companion, willing helper" (Book 1, p. 69). He is, indeed, a nightmare apotheosis of Brunner's Gerald Howson; born under Nazi bombardment to 15-year-old whore Rosa, mute, he rifles minds around him and commands his mother's obedience by a needle of cardiac pain. He has saved Barbara's telepathic ability from destruction, and captured her into devoted servitude.

What makes Dan Morgan's treatment of psi valuable, and psience fiction's frequent disabled characters finding a kind of solace or redemption via telepathy, is sympathy and empathy even for people so disfigured, reduced and embittered. Victor is imposing his will on an emotionally manipulated Barbara, one of the classic dreads the non-psychic entertain in face of sorcerous psi. We surmise that before the end of the quartet, Victor will be set free of his maimed shell and become a force for virtue, if perhaps by ethically ambiguous means. This happens in a surprising way, and by the end of the first volume.

And in the long arc of the series, told by different voices, we follow some characters from maturity to debility and death, a feature of traditional naturalistic novels not always seen in classic science fiction. Rebecca Shofield, in the opening book young and capable if hopelessly in love with Havenlake, is found in the fourth volume at the edge of death: "Becky was propped up by pillows at the head of the large, old-fashioned bed, a tiny, shrunken figure with snow-white hair and an uncompromising beaked face with vital coal-black eyes.... Becky was dying" (Book 4, p. 102). Like the Toby and Sid dyad, now transfigured, she awaits merging with the "mother-sea of consciousness" that is the unknown home country of the mind.

There are two problems that prevent this quartet from being one of the most impressive instances to date of psience fiction. One is the Gothic coloring of Victor's emergence as a monster of mind manipulation. But this is set aside in his death and transformation into a passenger in Havenlake's psi-deaf brain, loaning the parapsychologist his own ample power. The twins murmur gently in Barbara's freed and vengeful mind: "Such a tiny, helpless monster.... A poor Humpty, Dumpty creature, Barbara, can't you see?" Becky, opened to telepathy by the twins, agrees: "He was struggling for survival in the only way available to him. Should we judge him?" (Book 1, p. 153).

Less acceptable is the laughable attempt to describe these mental invasions and Victor's transfer from body to body. Granted, this was psience fiction written at what still amounted to the dawn of neuroscience, but we see the typical nonsense about "Taking into account the apparently unused cellular potential of the average human brain" (Book 3, p. 150), and an overly detailed parade of outdated information about the structure and operation of the brain. Victor initially plans to shift his mind into the exhausted cortex of Peter Moray:

> The personality of Peter Moray, defeated by Psi forces beyond its comprehension, lay helpless in retreat, imprisoned in a tiny complex of cells in one corner of the brain which had once been its kingdom.
> It was part of Victor's plan to exert all his terrible power and obliterate that identity completely, cauterizing it out of existence [Book 1, p. 146].

Luckily he had spent months studying every cell in Peter's body.

> Each individual impulse that made up the pulsating whole [of his "nebula of Psi energy"] had to be funneled into the host brain through the narrow channel of the medulla, and then established in permanent quarters within that brain....
> Deliberately splitting his awareness, so that it fissioned into a hundred tendrils of energy, each of which forced its own way along an individual nerve channel, he began the process [Book 1, p. 147].

As crude metaphor perhaps this suffices, but it suffers from what Arthur Koestler once called "the fallacy of misplaced concreteness."

Ψ

Putting such quibbles aside, the remaining three novels of the quartet maintain Morgan's generally solid sense of 1960s' British multiculture, even when pressed forward by a generation. The details, inevitably, are wrong, but as with Brunner's Telepathist in a world of woes, conflicts and a promise of healing, it is sufficiently credible even now to make the books readable.

In *The Several Minds* (a title drawn from William James: "There is a continuum of cosmic consciousness, against which our individuality builds but accidental fences, and into which our several minds plunge as into a mother-sea"), we learn that the human mind is divided into four levels, although this parsing is "a considerable oversimplification" (Book 2, p. 29).

The first level is the unverbalized stream of consciousness used by the psi-gifted for normal communication. The second is teleological reasoning, requiring close tuning between adepts. The third is perhaps Freud's Id, the boiling unconscious, choked with terrors and furies and nightmares and lusts, a dangerous place to enter and navigate. The fourth is "the basic, survival core of the individual" yet "innately in touch with some kind of cosmic consciousness, a common fund of knowledge that was the heritage" of the entire species, perhaps Jung's collective unconscious (Book 2, p. 30).

In a trope familiar from Bester and others, Havenlake risks a descent into the hell of the third level of his ferociously mad wife, Annette, first seduced by her and then bound, symbolically castrated and slashed in the arm by an imaginary sword. Victor, who now shares his brain, enters the dreadful parallel reality as a golden winged god, carrying his host away to the higher levels and then into ordinary awareness—only to find Havenlake's wounded arm bleeding freely. In a psi world, nightmares have waking consequences.

In the loathsome journalist Alec Glover, a sadistic sexual predator with powerful psi abilities he has not yet recognized, Morgan builds a counter to the earlier impression that psi must redeem even horrifying manipulators such as Victor. Intent on unlocking the secrets of the terminated British ESP program, Glover pursues Havenlake without success until he manages to force Annette to suicide. In despair and driving drunk, Havenlake leaves open a chink in his mind shield, and Victor seizes control, locking his host into the sort of small enclave to which he himself has been restricted.

Meanwhile, a young couple, Jerry and Sue Coleman lose their newborn baby to a brain embolism, and Victor chooses to move his mind to the dead infant, repairing the neural damage sufficiently to return the body to life. Learning this, Glover threatens to kill the baby unless the secret of serial immortality is revealed to him. Feeble as he is, the tiny relocated Victor manages a psychic modification of Glover's nervous system in a punishment-fits-the-crime fashion, connecting his orgasmic endocrine response to a fatal overload of his pituitary.

Ψ

In *Mind Trap*, 16-year-old Katie Mackinnon is an intelligent girl with the power to actualize fancies such a flying saucer landing in her slum school's playing field. Caught while stealing, she is committed to a reformatory for delinquents, and escapes by making herself invisible only to be threatened, before she fights back, with molestation by a truck driver. By a stroke of improbable luck (or is it careful psi guidance, always available in psience fiction?), she finds shelter in a nearby stately home owned by wealthy Henrietta Van Eps, with an Indian *guru* in residence who already knows her name.

Peter Moray, meanwhile, is conscripted by a cold-blooded counterintelligence agent to probe the mind of a profoundly neurotic scientist who is charged with providing crucial information on antimissile systems to "the eastern bloc." Denying his guilt, the man is slain by psychic means while meditating in his cell before Moray can get deeper into his mind than the first level. By another stroke of luck, it turns out that recently dead brains can be reactivated and probed by PK, yielding one confused blurt of data: *Saranamee*, an Indian or Sri Lankan name, perhaps a repeated mantra. The reader might surmise that a psi cult of meditators is the link between Katie, the Havenlake group now living in retreat, and a spy network. Actually the situation is less obvious than that, revealed by the ordeal of Maurice Ableson, a research biologist working on the production in huge quantities of a mutated virus, allegedly for "defensive" purposes but obviously a weapon of mass destruction.

Ableson, a Jew especially conscious of the horrors of genocide, is on the desperate edge of psychological collapse when he visits the Tahagatha Ananda School of Creative Meditation, scornful of the gullible dupes in attendance but falling quickly under the *guru*'s sway. Given a mantra to repeat, he enters into "a warm-womb darkness filled with the scent of roses, my consciousness ... in a state of blessed nothingness, a freedom from conscious thought and action" (Book 3, 109). But nightmare embraces him, the horror of a world devastated by his *Cirensis butor* virus. Emerging in utter horror, he is offered relief by the guru—if he allows young Katie, who has been working for Creative Meditation during the last year, to enter his mind and drain off these toxins.

So Morgan has introduced yet another kind of psychic ability in this third volume, while maintaining a degree of continuity in the quartet. Katie, of course, is drawing out a great deal more than mental poisons, archiving the secret research findings in a specially modified storage vault of her neural net. The novel tears along, changing narrators by turns, leaping over narrative and logical gaps, until at last an attempted flight from England is thwarted as Katie, Peter and Barbara perform upon the wicked Van Eps a kind of Besterian Demolition (Book 3, pp. 188-89):

> it was the stuff of sheer horror, the dreadful, unseen terrors that lurk in the unexplored parts of the mind ... bad as they might be for us, Henrietta was the one for whom they really meant something, because they were the personal myths on which her very existence was based.... These were the pressures, the veritable face of the Medusa ... beyond which there could be...
> Nothing ... nothing ... nothing.

Ψ

The quartet's finale, *The Country of the Mind*, moves ahead perhaps a quarter century, some 15 years beyond publication date. Victor's mind has long ago pervaded and repaired the dead baby, who has grown to adulthood. He is now Dr. Victor Coleman, psychiatrist, matured to adulthood, peace, and even love with the beautiful dark-skinned Flower. Britain by now, however, has become a seething witches' brew of bigotry and gang violence outbreaks, under a fascistic and racist Prime Minister. (In the real world, conservative Margaret Thatcher would become prime minister in 1979 and be gone by the end of 1990, when unemployment was down to about seven percent after a high of twelve percent six years earlier, and the population was still more than ninety percent white.) The UK sketched as background to this final volume is closer to today's troubled and fragmenting Britain.

Victor, at a party in Flower's parents' home, is threatened with castration or death by a hatred-filled, razor-wielding West Indian Rasta, but saved by another black man. Flower is gangbanged to death in a railway station by brutish *Clockwork Orange*-like Hobs in blue pinstriped uniform, as two "solid, respectable citizens" peer down, the woman's "mind behind the jeweled spectacles" reflecting with disdain: "*Only some black Immigrant whore getting what she deserves—*" (Book 4, p. 66). Rushing too late to her aid, Victor reverts to his earliest rage and exacts instant PK vengeance—but a sort of feedback shock robs him of his psi access. Black friends of Flower are arrested by a prejudiced, psychopathic cop and charged with her murder to save the reputation of the killers, one of them the scion of a wealthy power in the land. Becky's Association of psychics refuses to intervene, deeming it too soon to make themselves known publicly, lest this trigger a pogrom against the psi-gifted. It is a crisis implicit from the very beginning of the quartet. Even so, the final mystical uplift of Becky's postmortem communication, and return of Victor's powers, is oddly Disneylandish, especially given the hard-edged treatment of the corrupt policeman:

> *Then you died to save me.*
> *No, Victor, I died to save me—to escape from that worn-out shell, to become myself again...*
> Apparently alone on the mountain top he knelt ... as the four others of the inner council joined him in a Psi-communication deeper than he had ever before experienced.
> ...they gathered together on the Psi-plane and watched the proud image of Becky move away from them toward the distantly glowing nebula that was the gateway to the mother-sea of consciousness [Book 4, pp. 133–34].

More persuasive is thematically paralleled slow, painstaking PK repair by Victor of the hemorrhage-damaged brain of racist Prime Minister Dunleavy's wife, Ella. And the denouement is a painfully achieved moral choice when Jamaican guerrillas, led by a friend of Victor's and the late Flower, prepare to trigger a race-based defensive civil war. This is psience fictional ethics at its usual crossroads: use psi to transform another human into a puppet, at the cost of his sanity, or allow others to perish? It provides a suitable capstone to this flawed but ambitious tetralogy.

29.
Richard Cowper, *Breakthrough* (1967)

John Middleton Murry, Jr. (1926–2002), wrote science fiction under the name Richard Cowper, perhaps to avoid confusion with his father (1889–1957), once a famous British upper middle class literary critic, socialist, pacifist, Christian and finally conservative, thorn in his son's ambitions as a writer. Murry, Jr.'s fictive voice is lucid and coolly emotional yet romantic, having not very much in common with classical science fiction textuality, even H. G. Wells's. *Breakthrough* seems in some ways like a Noel Coward drawing room comedy, but with mythic undercurrents and overtones. It is clearly psience fiction, but as the online *Science Fiction Encyclopedia* notes: "Cowper does not usually link telepathy with the idea of the Superman, as is more normally found in U.S. sf uses of the convention; instead, it can be seen in his work as an analogue of "negative capability."

That is the poet Keats's notion of a quality characteristic of those who attain high achievements, especially in literature, by their capacity "of being in uncertainties, mysteries, doubts, without any irritable reaching after fact and reason." In some respects this is always one of the marks of an sf writer and reader, akin to Coleridge's celebrated "willing suspension of disbelief for the moment." In other respects, it might be seen as precisely the reverse, since sf is often regarded as crucially concerned with facts and rationality, even when the entities and properties it deploys are entirely unreal (there are, to the best of our knowledge, no time machines, Martians or human galactic empires).

In his early novel *Breakthrough*, the background ambience is the level-headed common sense of the Senior Common Room at a university rather less steeped in the classics than Brasenose College, Oxford (founded in 1509), where Murry, Jr., took his degree. In the early 1960s, his invented Hampton University is eager to turn out students fitted for a technological future, each a sort of C.P. Snowman. So it is not altogether startling that Jimmy Haverill, a new lecturer in Literature, runs into the American Dr. Frederick Wolfgang "Dumps" Dumpkenhoffer, who identifies himself as "Psychology, Parapsychology and Psi Effects—*inter alia*." Dumps's pure research is funded by a Liverhome Fellowship, a "full five-year research programme guaranteed right through," with a do-it-yourself lab in a dusty attic, an "electronics whizz-kid" and a capable young woman assistant who speaks in beautifully modulated and rather classy sentences. Dumps comes brilliantly alive, although the lovely student with whom rather dull, anorak-wearing Haverill is smitten (as he might have put it), the orphan Rachel Bernstein, is a sweet, delicate, fey creature with an occasional edge to her voice.

What makes this novel interesting from a psience fiction standpoint is the invasion into its lightly sketched academic setting (early postwar, it seems, rather than its alleged early to mid-1960s) by a kind of vacuum-tube computer experimental parapsychology blended with Rhinean card guessing. Dumps has the routine student guinea pigs put through their paces as psi participants trying to guess which card will be drawn next. Their choices are punched into cards and then checked for accuracy. Miss Bernstein is remarkable for her almost compete record of failure in this task, which strikes Dumps as impossible. "She got so many wrong her score was well below anything permissible. Short of scrapping the whole law of mathematical probability, I've got no option but to assume she's pulled a fast one."

Evidently he has never encountered the notion of "psi-missing," proposed by British psychical researcher S.G. Soal in the 1940s, which points out that the statistics of chance events leading to this kind of extreme negative result are just as evidential for a psi effect as is disproportionately hitting the correct cards.

He also can't have heard of "displacement," another of Soal's suggestions after he found that often the cards chosen were not randomly wrong but significantly often matched the card turned up next. Thus, the possibility of precognition entered the force-choice protocol. When Jimmy lightheartedly suggests that Miss Bernstein might be calling the numbers a week in advance, the parapsychologist is agog, and runs the analysis on that basis, finding that Rachel has indeed scored fantastically well—precognitively.

Dumps, additionally, trots out the nonsense about the unused brain and overstates the virtues of hypnosis:

> Why, do you realize, Haverill, even the most gifted of us probably use less than ten per cent of our own intellectual equipment? There are whole areas of the human brain that are absolute *terra incognita* as far as function is concerned. Under hypnosis you can recall virtually everything that ever happened to you—everything.

Still, we are given reasons to suppose that Dr. Dumpkenhoffer, also a psychiatrist, is an ace at this sort of fringe science. He and his very proper electronics whizz devise a series of brain scanners and transducers that search for the relevant neural structures in Miss Bernstein's head and also in Jimmy's, since he now seems to have entered into some kind of resonance with her.

When he dreams of something like a mythic landscape with arena, dais, green fire, and other impedimenta familiar from the dreamy art of Romantics, she shocks him by producing a newspaper clipping on the discovery of lost Hyphasis in Asia Minor, and a painting of exactly his dream scene, by a once famous artist: *Imaginary Landscape* by John Martin (1789–1854).[1]

He begins to hear her voice, in italics, in his mind, and eventually whole apparent messages:

> Does she understand what is afoot? "'Understand' isn't the right word.... It's— it's a bit like learning to swim. Knowing the strokes isn't enough in itself—you have to be able to let yourself go—to, well, to make an act of faith in yourself."

At length he receives a somewhat Atlantean vision in which he is addressed as Haalar (a version of Prometheus) and Rachel as Araaran; they are told by the voice of Kroton that they are the last of the Sky Children, that Los and Rinam are buried beneath the sands of Time, and so forth. A final vision is manifested, seen differently by all the principals, when Rachel goes under the mind-reading helmet. Somehow they are all *en rapport* with her. Jimmy raises his eyes to the display and

> I saw the whole curve of its surface subtly rearranging itself into a fantastic complex of shifting crystalline shapes; myriad upon myriad of light-splintering pinnacles and towers, eye-ravishing caverns and endlessly spiraling arcades, which shimmered and glanced and spun, mounting up and up, forever and ever, into a fathomless, dazzling sky—a firmament so unutterably immense that the heart shuddered to look upon it.
>
> I did not need Araaran to tell me what I was gazing upon; this was the palace of the Sky Children, the creation of ancient Kroton's prophetic dreams; something so far beyond the boundaries of material imagination that we three humans who gazed upon it were held in awestruck silence.... It was as if we were being shown some possible future that lay within ourselves and was attainable.... This or nothing was the secret of Los.... Rachel's message; this her revelation; this her home.

The electronics maven sees "Equations."
Dumps sees the face of God.
In due course Jimmy and Rachel marry, having put all this behind them with suitable British phlegm, Dumps literally disappears while trying to absorb Rachel's brain record directly into his own, they visit the lost city in Asia Minor, and the novel no longer seems to have much of the psience fictional in it; rather, it seems like a very mild Kingsley Amis attempt to emulate, with a straight face, a Rosicrucian or Theosophical fable. Still, it wins its place in any history of the genre, as does, in a slightly different way, Colin Wilson's mildly Lovecraftian *The Philosopher's Stone*, to which we shall return.

30.
Anne McCaffrey,
Talents Universe (1968–)

What if psi's chief evolved function is to work recursively and directly *on the genome itself*? A version of this not-quite-Lamarckian notion was published in an *Analog* science fiction story by Anne McCaffrey, "A Womanly Talent," later expanded into a series beginning with *To Ride Pegasus* (1973) and continued in *Pegasus in Flight* (1990), *Pegasus in Space* (2000), and beyond. The parapsychic Talented began to emerge in the twentieth century and were brought together and validated with the use of "Gooseggs" (a form of "ultra-sensitive electroencephalograms," which have almost nothing in common with the real world Global Consciousness Project's much later EGGs, or ElectroGaiaGrams).

The sequence, somewhat like Dan Morgan's, maps a future history of the psi-gifted, and eventually continues beyond *Pegasus* to *The Rowan* (a deeper interstellar future) and many others. It begins with a serious car accident that the precognitive Henry Darrow fails to avert—rather unbelievably—because he is distracted. Perhaps destiny is lending a hand, because he thus meets and instantly falls in love with intensive care nurse Molly Mahony, who has a paranormal healing ability. Together they search out and gather highly effective psi operators—telepaths, precognizers, clairvoyants, telekinetics and teleporters (McCaffrey's misleading term for a levitator), empaths, finders, pyromaniacs who are attracted to imminent fires without causing them but often found guilty of arson.

In the late 20th century, these strays found the first Center for Parapsychics in Jerhattan (presumably Manhattan after it absorbs New Jersey) where people use credits rather than dollars and IBM cards for data storage. Darrow charges plenty to government and law enforcement for the hire of his operatives, and gains protection as well as funding. Inevitably, once psi is out in the open as truly more than trickery and self-delusion, the non-psychic majority cringe in resentful fright, and graffiti slogans and worse begin to blight the psychics' lives. All of this history is presented with some narrative skill, working by ingenious and sometimes moving ways through the possible plot combinations familiar from the corpus of psience fiction.

In "A Womanly Talent," which forms one seed of this sequence, young Ruth Horvath, apparently without any psi Talent, has recently married powerful precognitive Lajos Horvath, a fire hazard sensitive who advises an insurance company. During sex, remote monitoring instruments reveal in their "coital graphs ... a tight, intense, obviously kinetic pattern." So why and where is Ruth expending this prodigious kinetic energy?

The answer surprises everyone in the story, but tends to strike one these days as at least somewhat sexist: "For the exercise of a very womanly talent.... What is the fundamental purpose of intercourse between [sic; she means "with"] members of the opposite sex?" Well, reproduction (leaving aside all the pleasurable and social benefits). With this clue, the Talented discover that Ruth's new baby Dorotea manifests inheritance patterns that are genetically possible *but extremely unlikely,* making the baby an immensely strong telepath. So what Ruth did, completely unconsciously, was to rearrange by psychokinesis "the protein components of the chromosome pairs which serve as gene locks and took the blue-eyed genes and blonde-haired ones out of cell storage. And what ever else she wanted to create Dorotea." She does the same in order to perform a miracle of healing on someone else. "She can actually unlock the genes!" cries one amazed specialist.

It was a provocative idea, when generalized (and only a decade and a half after Watson and Crick first identified DNA's structure and code). Could psi, at its deepest level, be a process for optimizing each species' genome as swiftly as possible within a stable environment that precognitive psi itself suggests will remain unchanged for many generations? Perhaps this is a clue to the somewhat surprising discontinuities in evolutionary history claimed by Stephen Jay Gould and his colleagues, dubbed "punctuated equilibrium." Species can appear convulsively (on the geological timescale, that is), and then persist almost unchanged except at the micro level for many millions of years, until a new challenge, often environmental and catastrophic, leads to a new leap.

Might it be that psi acts to consolidate such major adaptations, even to feel out potential pathways in evolutionary space and encourage the structure of the genome to settle into a maximally appropriate form? A version of this idea forms the basis of a radical theory—quantum evolution—suggested by a professor of molecular genetics, Dr. Johnjoe McFadden, and by mathematical physicist and Templeton Award winner, Dr. Paul Davies.

Imagine something like a genetic sequence that manipulated and rewrote its own alleles directly. Such an effect could transform the typical allele into a psi-conducing configuration. The presence of such genes might then encourage a competing selection of the original genes that defend the "non-psychic" alleles, acting to suppress or obliterate the "psychic" form. Paradoxically, they might even do that by something like psychokinesis, conscripting the talent they act to suppress.

Presumably such "prionic psionic" effects would eventually attain equilibrium, either becoming self-limiting or reaching a balance by mutual contest. If not, a species might fairly quickly tend to become saturated with one form or the other. More likely, some proportion of the population might use a mixed strategy that is not only permissible but might well have a decisive advantage.

Ψ

In the extreme, might psychic superstars be parasites powered by the future psychic capacities of their descendants (and perhaps by the abilities of their long-dead ancestors)? Presumably such psychic vampirism would have to be limited by some law of diminishing returns in either temporal direction. It pays the family line to be very fertile, and for all gifted individuals to exercise their own fecundity. On the other hand, it risks various deleterious side effects, the sorts of accumulated disorders that are common within inbreeding populations. (Unless such genetic errors are self-repairing via PK, as in McCaffrey's universe.) This dynamic might seem to fly in the face of celibate saints and mystics, but it is notable that their *siblings* have traditionally been encouraged to be hyper-fertile.

Suppose that some "witch burner" gene conduces to various kinds of furtive or masked opposition to those carrying "psi genes." Those expressing this gene in behavior might not even be aware of their own gene-biased motives, in much the way that effective liars first convince themselves that they are speaking the truth, thus defeating the inbuilt lie detectors of their victims. There is no *a priori* optimal outcome in evolution, only a nonstop tussle between existing competitors and the genetic material (including rare mutations) available for organisms to build into their offspring. The giraffe doesn't know that a longer neck would favor its child, even if it knew how to bring that about by Lamarckian onboard genetic engineering—*unless* there is a way to gain advance access to various (possible) future outcomes.

And as McCaffrey implied, that might be the key function of psi: to search the many pathways opening from this moment into the future, using precognition; to test any solution found against those discovered in competing genomes, both here and in alternative realities, by clairvoyance or remote sensing; and finally, to tweak the genome of which it is one expression, by anomalous perturbation (in this circumstance of literally mindless cells, "*psychokinesis*" is a particularly ill-formed term).

Indeed, it is possible that the origin of life itself forms part of a "psychic" temporal loop, with conditions in the future reaching backward into the multiple possibilities available in the past and biasing those conducive to its own existence. Is this more preposterous than the standard explanation for the origin of the universe: that it quantum-tunneled from a state of nothingness into being, *just because it could*? With psi thrown in as an extra empirical factor currently ignored by almost all scientists, decisive breakthroughs might be anticipated.

Ψ

The Talents universe explores some of these implications of psi reconfiguring its own genetic and phenotypic carriers, or at the very least allowing them to compensate for disabilities. *Pegasus in Flight* features two Talented children. One is Tirla, a twelve-year-old illegal from the ghastly Linear overpopulated slums who should never have been born, but survives to become a key telepathic translator helping bond competing cultures. The other is Peter Reidinger, a paralyzed teen who uses psychokinesis to move both his body and heavy materials directly by mental intention, and who returns in *Pegasus in Space* after learning to connect his psi abilities to power generating machinery crucial for space flight.

The Rowan centers on an infant girl who survives the destruction of a mining community and develops into a staggeringly powerful psi Prime who eventually commands the interstellar Federal Telepath & Teleport network, moving information and shipping instantaneously from world to world. These psi gifts are now calibrated between T1, the very powerful, through T12, the least strong. There are threats from evil aliens that only the psi-gifted can defeat; there are romantic opportunities of the kind McCaffrey specialized in. These are the repeated tropes of psience fiction: outsiders spurned by the narrow and conventional, finding independence and achievement not just despite the obstacles of their deviations from the norm but through them. It is no accident that these books are usually marketed as YA, and embraced by the lonely in their loneliest years. This is psience fiction as compensation for the slings and arrows of outrageous fortune, and in high wish-fulfillment mode, effective because of the imaginative detail and pathos McCaffrey brought to her task.

31.
Philip K. Dick,
Ubik (1969)

If a single writer of fantastika deserves to be nominated as the prime creative genius of psience fiction, it has to be the late Philip Kindred Dick (1928–82). His psi-inflected work was rarely constrained by what parapsychologists have learned about the paranormal, yet his verbal texture adapts J.B. Rhine's lexicon (especially "precogs" who see and sometimes manipulate the future). His stories take the form of futuristic adventure, yet in the main his psience fiction is both epistemological (concerned with what we can know) and ontological (exploring what *is* reality, in particular what is *human*). It frequently has a spiritual resonance yet it is never doctrinal.

31. Philip K. Dick

Perhaps the most curious aspect of his psience fiction career is Dick's failure to attract the support of John Campbell, who was at that very time promoting psi as a major sf ingredient, and one based on reality not sheer fancy. Dick spoke with irritation of the pressure Campbell exerted on his writers:

> In the early Fifties much American science fiction dealt with human mutants and their glorious super-powers and super-faculties by which they would presently lead mankind to a higher state of existence, a sort of Promised Land. [Campbell] demanded that the stories he bought deal with such wonderful mutants, and he also insisted that the mutants be shown as (1) good; and (2) firmly in charge.[1]

It is also notable that Campbell only ever purchased a single Dick story, "Imposter" (*Astounding*, June 1953); he had a specific animosity toward Phil Dick's approach. Dick noted in a comment on a later story:

> Horace Gold at *Galaxy* liked my writing whereas John W. Campbell, Jr. at *Astounding* considered my writing not only worthless but as he put it, "Nuts." By and large I liked reading *Galaxy* because it had the broadest range of ideas, venturing into the soft sciences such as sociology and psychology, at a time when Campbell (as he once wrote me!) considered psionics a necessary premise for science fiction. Also, Campbell said, the psionic character in the story had to be in charge of what was going on. So *Galaxy* provided a latitude which *Astounding* did not.[2]

His plots range from hasty to the point of deranged, and random through to ingeniously logic, and can at the same time be hilariously funny. Before we look closely at *Ubik*, the kind of dizzying narrative that Campbell might have cited to justify his assertion that Dick was *nuts*, we should consider some of the more explicitly psience fictional shorter works. Three are characteristic, yet different from each other in their use of psi: "A World of Talent" (*Galaxy*, 1954), the lumpishly editor-retitled "Psi Man Heal My Child!" (*Imaginative Tales*, 1955), and "The Minority Report," *Fantastic Universe*, 1956—filmed with many changes by Stephen Spielberg as *Minority Report* (2002), then sequeled into a swiftly truncated television series in 2015.

Ψ

In "A World of Talent" a small settlement of humans on Prox III (a planet of the closest star to Earth, Proxima Centauri) is divided into Norms, or traditional *Homo sapiens*, Psis of various kinds (precogs, Parakineticists, Telepaths—who rules the roost—Animators, Resurrectors, etc.), and Mutes or mutants. Eight-year-old Tim resembles a number of the damaged or variantly abled autistic children in other Dick novels and stories. Tim is troubled by Others, of both the Right and Left kind; these hide in dark places. His mother, Julia, is a talented precog; his father, Curt, is also precognitive, but to a less marked degree. At a party, the psychic parents foresee that the night will end in a bad fight, but they decline to do anything to avoid this unpleasantness. This is a choice; there is no absolute predestination, for otherwise there would be no advantage, personal nor evolutionary, to the precog gift.

The argument arises from Curt's liaison with young Pat Connley of Prox VI, fetched to Prox III by Big Noodle, an obese idiot with extreme powers. Pat proves to be the first of a new class of Psis, the Anti-Psi, who can block or interdict the usual classes. When Pat is murdered at the instigation of Reynolds, chief of the telepathic Corps, the young teen Sally who in effect controls Big Noodle goes too far: the obese idiot kills her but is struck before she dies by a "space-transformation": "The vast body had become a mass of crawling spiders" (*Collected Stories*, Vol. 3, p. 347).

The child Tim, now a mid-thirties man and master of movement in time as well as space, reveals that the Right Others have sponsored the emergence of an Anti-Psi. "I could wipe out everything that stands. I precede everyone and everything.... I am always there first. I have always been there" (p. 351). And he places his father on the single time path of the multiple realities on the board where Pat is still alive, and in his arms.

This trope of a paranormal power to move back and forth in time is at the heart of "Psi Man Heal My Child!" which presents a grim post-nuclear future in which a small gathering of the psi-gifted attempt to maintain themselves while helping heal the blighted normals. Jack Tremaine is a member of the psi Guild, moving back and forth along his own time line from adolescence to an aged 71, trying to retroactively prevent the apocalyptic war that has already ruined the planet. Many times he has approached General Butterford just before the bombs fall, carrying a parcel of Butterford's own bones. On this twelfth attempt, he has no better luck in piercing the military mind-set that always triggers the war on all sides of the conflict. He is slain by Stephen, a crazed psi with fascist instincts, leaving four members of the Guild locked in a stand-off. It is not a satisfying adventure story finale, but it is the truth of Philip K. Dick.

"The Minority Report" explored with more rigor the consequences of precognition and the use of that knowledge to manipulate the future—in this case, by arresting all potential criminals just in advance of their crimes, up to and including murder. The driving irony here is that "bald, fat and old" John Anderton, the inventor and developer of the mechanisms enabling the prophylactic Precrime Agency, is now Commissioner of Police—and finds an analysis report that projects his imminent murder of Retired General Leopold Kaplan, Army of the Federated Westbloc Alliance. He not only has no such intention, but had never heard of Kaplan until this forecast. Is he the subject of a monstrous conspiracy, perhaps including his own attractive wife, Lisa, and would-be replacement, Ed Witwer? (It turns out to be far more complicated.)

Dick's presentation of the three stunted precogs whose nonsensical prattling yields these reports is horrifying, one of the most chilling images in all psience fiction.

> In the gloomy half-darkness the three idiots sat babbling. Every incoherent utterance, every random syllable, was analyzed, compared, reassembled in the form of visual symbols, transcribed.... All day long the idiots babbled, imprisoned in their special high-backed chairs, held in one rigid position by metal bands, and bundles of wiring, clamps. Their physical needs were taken care of automatically. They had no spiritual needs.... Their minds were dull, confused, lost in shadows [*Collected Stories*, Vol. 4, p. 73].

Unlike most psi-gifted figures in Dick's fiction, the horrid secret is that "The talent absorbs everything; the esp-lobe shrivels the balance of the frontal area. But what do we care? We get their prophecies" (p. 73). It is an interesting prefiguring of the fate of Gerald Howson in Brunner's *Telepathist*, but far more terrible. There is no such known neural correlation in real paranormal studies of the kind developed by the Star Gate program or the Princeton Engineering Anomalies Research unit, but there have been claims that some autistic children manifest unusual psi behavior.[3]

The final irony of the story is Anderton's discovery that all three reports on his supposed future crime differ in crucial elements, since each is out of phase with the others, drawn from an alternative future cosmos, like the sigma sequence in Blish's *Jack of Eagles*. The earliest report foresees his murder of Kaplan; the second is informed by his awareness of the previous report, so he chooses not to commit the murder; the final report takes account of his change of mind and allows him to resolve his doubts by choosing again (p. 99). Rather than undermine his life's work in Precrime by raising crucial doubts, he very publicly kills Kaplan and allows himself to be taken into custody by police and charged with murder on the basis of the bogus "majority report."

Ψ

Many of the exuberant and perhaps amphetamine-stimulated Dick novels of the 1960s contain elements of psience fiction, but especially of metaphysical inquiry enacted in the *mise en scène*. As Polish polymath Stanislaw Lem pointed out, phantasmagoric fiction, in the vein of Kafka's, takes us inside the warped allegorical perceptions of an outsider. But with most of Dick's, the flux of change is in "reality," while the observing characters maintain their solidity, even as what they see and feel stresses them nearly to the point of frank insanity (Lem, 1985, pp. 106–135). This literal deconstruction of the text of the world in *Ubik* takes the process about as far as it can go, emblematized by the titular ontological spray or goo that can reverse aging and other deterioration (when used, as we are repeatedly and amusingly advised, "only as directed").

His gift for naming is brilliantly off beat. *Ubik* opens with news that "the top telepath in the Sol System fell off the map" while pursued by three antipsi "inertials" employed by Runciter Associates in New York. This psychic, S. Dole Melipone, vanished in a motel, the Bonds of Erotic Polymorphic Experience. A page later, the Beloved Brethren Moratorium is being opened by its owner, Herbert Schonheit von Vogelsang. A Moratorium, we find at once, is not a kind of temporary prohibition during which a particular activity is forbidden, but rather a place where the remains of the dead are warehoused in a kind of arrested mortality, available for conversations with the living— but that, too, of course, is indeed a sort of uncanny moratorium on decay, as well as a kind of mortuary.

Besides, the dead are not exactly defunct, like corpses stored at very low temperatures for indefinite cryonic suspension awaiting revival made possible by future technology. People dwell in "half-life" where communication can continue between these quasi-dead and the living, although at the cost of draining vitality. (Note that the term "half-life" is borrowed from radioactive decay, a hot topic in the 1950s and 1960s when everyone dreaded and even expected nuclear exchanges between the military superpowers at any moment.)

Despite this initial emphasis on telepaths and their opponents, much of the strange communication and epiphanies in the novel seems to be conducted electronically, as between the half-life deceased who appear to share a virtual reality with other dead people. Are the instruments undergirding this technology of almost miraculous contact just a form of what we might think of as Tweets or Instant Message or even smartphone links, or are they psi machines of a post–Campbellian kind? Or both, blended? In any event, it seems appropriate to regard much of Dick's work as psience fiction, instantiated here in a cloud of unknowing pierced by a hard-edged, down to earth and even pungent solidity of character. Even so, the name of the main character, Joe Chip, evokes the possibility that he is a kind of integrated circuit or "microchip," a computational device well established by the time Dick was writing the book.

The plot is complex and self-questioning, which is all we need to know for our current purposes. Glenn Runciter and his half-life wife, Ella, run a business that protects other companies from spying and attacks by psi operators. Joe Chip is one of his employees. Runciter and many of his staff are lured to the Moon where a bomb apparently kills Runciter. Chip and the other staffer return him to Earth for preservation in cold-pac, alongside Ella.

But the world turns strange. Everything is aging around them, diminishing, halting in a composite of 1939. Runciter's face appears on their banknotes.

One of the inertials, Pat Conley (whose doppelganger appeared as antipsi Pat Connley in "A World of Talent"), has the psi power of changing the past. The household product Ubik, a domesticated Eucharist or dime store theophany, can be used as a spray to reverse history's fatal entropic decay. Lem describes it as a "canned Absolute," the result of the mutual contamination of styles of thought from two different ages, "the incarnation of abstraction in the guise of a concrete object" (Lem, 133–34). Joe Chip realizes that he and the other staff probably died on the Moon and remain there, under life-sucking attack by a demonic being named Jory Miller.

Yet from Runciter's perspective, this is equally doubtful, especially after Joe's face appears on a fifty-cent coin. Ubik finally announces itself as the creator: "*I made the suns. I made the worlds. I created the lives and places they inhabit.... I am, I shall always be*" (p. 190). This is arguably psi pressed to the point of the divine, the ontological primary, proving that in his itchy, provoking way Phil K. Dick was the true master of zany psience fiction.

32.
Colin Wilson,
The Philosopher's Stone (1969)

The fabled Philosopher's Stone (prior to the first volume of the Harry Potter saga and its retitling for Americans as *Harry Potter and the Sorcerer's Stone*, which means nothing at all) was an alchemical cure for mortality. This is the titular sense of pop-polymath Colin Wilson's best science fiction novel, a book avowedly about the quest for a specific effective against inevitable death. It shares affinities with Wilson's earlier novel *The Mind Parasites* (1967), which he saw as a play on the horror themes of Howard Lovecraft but which more exactly seems a phenomenological retread of Eric Frank Russell's *Sinister Barrier*, a 1939 serial in John Campbell's fantasy magazine *Unknown*. In *The Mind Parasites*, it is discovered that humankind has always been infested by invisible creatures that, vampire-like, suck the life force out of us cattle, leading inevitably to angst, weariness, narrowed consciousness and finally death.

In *The Philosopher's Stone* a range of paranormal abilities is uncovered after the injection into a patient's cranium of tiny fragments of a handy soft metal, Neumann's alloy. This blend of metals with a touch of graphite has a "delayed action effect," absorbing a current and releasing it, augmented, with

The Philosopher's Stone (1969)   139

energizing results. Our narrator is Howard Lester, who with his sponsor Sir Henry Littleway tries this on himself as a means of expanding his consciousness, rather as LSD and psilocybin were claimed to do around the time the book was written. It makes a new man of him, and curiously this seems to have influenced the writers of the back jacket copy on both UK and U.S. editions, which both dub Lester "Howard Newman." Perhaps that was Wilson's original name for the man of enhanced paranormal powers, until he changed it to avoid painful obviousness.

Lester is clearly a wish fulfilling image of Wilson himself, had he managed to find an imaginative, wealthy and well-born patron like Sir Alastair Lyell, who met Lester as a poor lad of thirteen and took the boy under his wing. When Lyell died in 1967, he left a substantial income to Howard, who by then was working on *Principles of Microbiology*. Some years of study later, Lester hears on the radio a Leath lecture (that is, a cousin of the very prestigious BBC Reith lecture) by Sir Henry Littleway who argues that "Man's evolution has been the steady growth of his independence from the body and the physical world" (p. 38). Soon Littleway is recommending major phenomenologists: Husserl, Scheler, Cantil, Merleau Ponty, Leicester (p. 40). ("Leicester," pronounced "Lester," is, we understand, Wilson himself, in his existential avatar as author of *The Outsider, Introduction to the New Existentialism*, etc.) Henry sends Lester a copy of *Ageing, and the Value Experience*, by Aaron Marks. (Marks, we understand, is Abraham Maslow, who in our world named the "peak experience," a condition of clarified consciousness of exactly the kind Wilson sought and promoted all his life.) Soon they are fast friends, intent on accelerating the evolution of mind, and via that development, *inter alia*, to put an end to death.

Littleway and a colleague perform a trial experiment on Howard, inserting two electrodes of Neumann's alloy, and within minutes he is transformed:

"What are you experiencing?"
"Ordinary consciousness."
He looked surprised and disappointed.
"You mean there's no change?"
"Oh yes, there's a change. I didn't say it was *everyday* consciousness. It's everyday consciousness that's sub-normal. This is normal" [p. 98].

In Part Two of the novel, a full-blown examination of this new-man variety of awareness, will and mental activity proceeds at increasing velocity. This is where Colin Wilson swerves into the domain that in two years' time he would go on to explore in several rather large volumes of popular parascience, notably *The Occult* (1971), with the New Age front cover subtitle (surely not Wilson's choice) *The ultimate book for those who would walk with the Gods*, and its sequels *Mysteries* (1978) and *Beyond the Occult* (1988). In *Mysteries*, he concludes that

32. Colin Wilson

> Human beings will one day recognize, beyond all possibility of doubt, that consciousness *is* freedom.... Instead of wasting most of its energies in retreats and uncertainties and excursions into blind alleys, consciousness will recycle its energies into its own evolution. The feedback point will mark a new stage in the history of the planet earth.
>
> When that happens, the first fully human being will be born [*Mysteries*, p. 630].

The mystical assumption here, contrary to the detailed findings of science, is that evolution has a goal, a kind of theodicy, rather than a history. We have seen this premise advanced with varying degrees of confidence in a number of exemplary novels of psience fiction. One difficulty in reading Wilson's optimistic and encouraging novels, which are propelled by ideas, is how very silly so many of these ideas are, how foolishly his characters rush into blind alleys. Since the novels are fiction, this is no great crime, any more than a romp in the Hollow Earth is felonious when advanced by the scientist Rudy Rucker, or time machines and faster than light starships in any number of canonical sf works. The problem arises because we are constantly made aware that Colin Wilson means what he says, and patently regards himself as a great scholar and philosopher. It was astonishing to read in a review of *Mysteries*, for example, by the notable novelist Christopher J. Koch, that "here is an adventure in thought with a mind whose penetration, good humor and self-awareness betray no hint of the crackpot" (*Sydney Morning Herald*, 3 October 1979). *Au contraire*—the headlong chasing after false gods (if one may borrow the metaphor and apply it to truth claims) is dismaying.

Even so, *The Philosopher's Stone* is psience fiction, not parascience, not even hard core science fiction, so we can lean back and take it in with as many grains of salt as necessary. What is remarkable is how Wilson's magpie attraction to every new or old shiny or rusted idea sometimes leads him to some astonishing insights. Lester discovers that he and then later Littleway have developed "time vision," which in some sense is just what clairvoyants have always claimed but that nowadays seems a notable precognition of remote viewing, which would not be formalized until the early 1970s.

Visiting the aged governess of Littleway's late wife, near Stratford on Avon, they take tea on the lawn of an old Tudor house. Howard drowses, musing on the ghost of a lady in blue carrying a small black dog. He notes a shallow ditch that, four centuries earlier, might have been a moat—and there it was. He did not *see* it filled with water, but imagined it "*as if in a dream*" (p. 113), then imagined the lady in blue coming from a side door, the dog running ahead of her toward the old bridge across the moat. All this proves historically and unexpectedly correct. What Howard has done closely resembles the process used by expert military remote viewers such as Pat Price and Joe McMoneagle, who sketched scenes (without, in their case, knowing the

The Philosopher's Stone (1969) 141

location in advance) that were sometimes detailed to the point of instant, undeniable recognition by judges, yet containing elements from the past that no longer existed.[1]

Something similar happens when the two visit a poltergeist site at the Old Rectory, invited by the expert Sir Arnold Dingwall. (Like some of the other characters, his name is drawn from those who have studied the paranormal, in this case the skeptic E.J. Dingwall.) Dull husband, stylish and bored second wife, two daughters and a would-be biker son live in what they think is a building some doors from the now demolished Vicarage, in the 1860s the scene of a puzzling poisoning. Howard locks onto that murder, and knows at once that the scene was actually the Rectory, now subject to loud crashes and thuds and the usual panoply of poltergeist activity. He understands that this is a manifestation of the spiritually deadened ambience of the family. When the flinging and wailing and piano banging starts up, he amplifies it by sheer intention and will, driving it in cycles, then carefully mutes it to silence.

But where had the raw energy come from to power this "noisy ghost"? Well, "the earth spins through space at a speed greater than an express train, and we ourselves and the chairs we sat in and the walls around us were masses of buzzing atoms. The air was full of every kind of wave and energy.... The core of the poltergeist activity was the *negativeness* of the human energies in this house" [156].

Howard Lester's own energies and capacity to direct them increase. He *imagines*, in this new sense of the word, scenes of Shakespeare's playhouse, and Goethe in discussion, and is soon penetrating deep time, using ancient pots and other artifacts to guide him to the period of their making and use. Some scenes are horrific; eventually he is aware of Mu as it fell because, this being a Colin Wilson novel, "in the sky hung an enormous moon, blueywhite in color, that counteracted the earth's gravity and caused the tremendous growth of all the living creatures of Mu" (p. 245), and its great all-but-immortal magician king K'tholo (Lovecraft's Cthulhu, much humanized), and farther back to the Great Old Ones who morphed the genes of beasts to create humans ("as told in the Vatican Codex" (p. 256). The convulsions in the Earth caused by the arrival and departure of various moons and comets have a Velikovskian rather than psience fictional flavor, but it is clear that the suppressed powers of the ancients were psionic.

Finally, Howard Lester meets a beautiful young pregnant woman with two adorable children and marries her. They share a profound telepathic bond, and as she types his memoir

> her prefrontal cortex is beginning to work of itself.... Awaken[ing] in her the hidden evolutionary drives. And I suspect they may be transmitted directly to the child that is due to arrive in a few weeks. If I am right ... [p]ower over the

prefrontal cortex would be transmitted by heredity, and it would only be a matter of time before the whole human race had developed that power [p. 268].

You can't ask for more than that, however unscientific it is, from psience fiction.

33.
Joanna Russ,
And Chaos Died (1970)

Here, in Joanna Russ's second novel, is the best psience fiction novel in English and perhaps the world, despite its passages and implicit arguments that make 21st century sensitives squirm. The language is muscular and exquisite, but more than that it is wonderfully *thought*. Despite its convolutions and narrative quirks, nothing is here on the page for the sake of plodding forward. It is dance, in more than four dimensions. This is surely why the wordmeister Fritz Leiber said in his back jacket quote for the Ace Special paperback original:

> *And Chaos Died* explores more fully than I have ever seen what telepathy and clairvoyance would actually *feel like*. The result is a stunning achievement. There is a science fictional rationale for all the startling, stinging, and shocking imagery … and there is also a good, strong, human story, well resolved.

Is the story *well resolved*? Nearly half a century later, it disturbs some readers that marooned Jai Vedh quickly announces himself, when pressed by a vulgar macho oaf, the Captain, as a homosexual. In 1970, the shock was just that: his declaration that he disliked women and liked men sexually. Today, it's that he used the term "homosexual" instead of "gay," although of course not many people in 1970, even gay people, knew that "gay" was to be the only allowable term, other than "queer" or some challenging formulation including the word "gender."[1] This solecism, however, is not the gravest affront from lesbian feminist Russ: that would be her gay character's "healing," via plentiful energetic and mind-melding sex with a woman intent on setting him straight, and his acceptance of this change. So today's readers are likely to come to the text with preconceptions radically at odds with those of Russ (born 1937, died 2011), in 1970 soon to be author of that feminist classic *The Female Man* (1975).

On that same back jacket, black gay sf writer Samuel R. Delany praised the novel in *phenomenological* terms aligned with Leiber's: "Many novels have dealt speculatively with psi-phenomena, describing the effects on people and society.... Russ has taken it on herself to put the reader *through* the experience. She is wholly successful. *And Chaos Died* is a spectacular experience to undergo."[2]

The ellipsis above replaces the very incorrect word Delany used all those decades ago: *Miss* Russ. Later editions changed that to *Ms.* Russ. (She became a full professor fourteen years later.) The point being made here is not about Chip Delany's historically situated choice of words, but the way this problematic (as academics dub it, meaning "a problem") is a sort of synecdoche (as academics dub it, meaning "part used as shorthand for the whole") of the novel's flight path.

Psi, it is revealed, through living it, absorbing it, fighting with it, using it to fight with, is rather like this kind of transformative verbal passage from "raw" experience to clumsy word-token to sophisticated "cooked" word-token to elaborated and distilled experience. And all of this psi-mediated experience is rendered by Russ, in the paradox of written art, as dancing, enthralling word-tokens, because we readers can't, alas, assimilate it directly via ESP.

Consider this early moment of verbal music as physical dance and vice versa (Russ had studied under Vladimir Nabokov, and he is co-dedicatee of the novel):

> The Captain, whose face said *I must stop, stop me*, put one terrified hand under the doll's breast and another on its belly; Jai hooked one leg under the man's knee and brought him down three yards to the side; he knelt efficiently on the bigger man's back and twisted both his arms.
>
> Ah, good! Lovely! said the clearing, full of eyes. He let the Captain go. The big man stood up, brushed himself off, ran one hand over his hair and folded his arms severely. "What's the matter with you, you don't look well," said the Captain simply, and then his eyebrows went up a fraction as the meditating woman opened her eyes, got up abruptly, and casually stripped herself. She hung the violated dress on the branch of a tree. "I'm tired of this dress," she announced off-handedly. "I'm going to get me a new one."
>
> "My friend will make me a real Coco Chanel," she said.
>
> "—eel oh oh ah Nell"
>
> "veil as well," she said, "Come on," and, nude, she stepped easily out of the clearing, all moving buttocks and knees, each side a balancing line to the armpit, feet like hands or limpets holding on to the turf, and swaying ankles....
>
> "Oh, it's real enough!" said someone out loud and then into his ear, intimately, making his head swim in a rapture of mischief, *But what a dream, what a drama! You eye people, you're unbelievable!*" [pp. 16–17].

The plot, hidden within the convoluted prose, is simplicity itself. On the larger scale, we are shown in action two alternative kinds of social order. Two men from the home world find themselves on another with wildly different values. At the smaller scale, we learn what happens when a sensitive, nearly psychotic man from a damaged culture *learns better*. The men are rescued by a cruise liner and return to Earth, bringing with them a psi-capable woman, Evne, who teleports ahead of them and intervenes in ways that precipitate tremendous disruption.

Earth is at the end of its authoritarian tether, massively overpopulated by hyperstressed people with nothing to do but express self-destructive rage in ever more lurid *autos-da-fé*. You can buy lethal weapons from handy vending machines. On the eco-balanced world where Jai Vedh and the Captain crashed, a small population of what amounts to posthumanity has gained a range of psi powers. Children speak, but adults exist mostly without oral communication, in a state of Noöspheric but playful gestalt. Both communities seem to prize sexual exuberance, although the home planet's is sadomasochistic and deviant (Jai himself more or less rapes a "feeble-minded girl," and his 14-year-old sidekick Ivat desires sex with children), and the new world is far more civilized although apparently scornful of same-sex eroticism.

I suspect that neither of these worlds scored high on Russ's own value scale (circa 1970), and that she was mocking both extremes. This reading cannot easily be validated from the text. We have been led to accept by the extensive psience fiction megatext that communities of psions must be unusually virtuous and other-directed yet creative, because the other is, in a real sense, oneself. Since sf is basically an exotic adventure format, inevitably there are dangerous and even terrifying exceptions to this benign premise, providing thrills and chases necessary to adventure fiction, but perhaps any construction of psi-based societies as necessarily akin to one of Sturgeon's redemptive hive minds is also under scarifying inspection by Russ. It is a risky error, after all, to conflate the views and beliefs of a text's leading characters with those of its author.

What is especially extraordinary about Russ's experiment in psience fiction is its edgy hallucinatory stylistics, an evocation of other ways of seeing and manipulating everyday reality as well as realities of a future Earth utterly broken from the constraints and opportunities of any natural order. Psi in this novel provides access to the "thoughts" of rocks and grass as well as one's conspecifics. At the same time, the usual geographical boundaries known from pre-antiquity by rootless wanderers are abolished because people can float free of gravity, teleport themselves at will to distant places, see those places anyway without bothering with the trip, engage with others' minds and experiences.

In the rather bitter Chinese parable Russ cites as her chief epigraph, which gives rise to her novel's title, we are told: "The eye is a menace to clear sight, the ear is a menace to subtle hearing, the mind is a menace to wisdom, every organ of the senses to its own capacity" (p. 5). If this is true, then acute sensory perception and action and astute abstract analysis are not only dangerous but ruinous. Drilling through the unified primordial self simply murders whatever clarified understanding one wishes to set free. Is this the message of a world shaped by psi, sometimes presented as the operation of pure consciousness untainted by the empirical? Is it even the message of the Taoist document Chuang Tzu, from which Russ drew her citation? Perhaps not: her second epigraph presents an almost Campbellian "psionic machine" proposition: "There is a point beyond which you can't go without the aid of the machine ... there is a limit to how loud you can shout. After that, you have to get yourself an amplifier" (p. 5).[3]

Psionics, Evne tells the assembled leaders of earth, is just perception and education. It is not radiation. She mentions a fable: Inside is Outside, and Outside is Inside. "Action at a distance.... Any system of organization must be tied to an organic body, so there are limits.... Space, time and mass" (p. 182). It is revealed finally that the people of the nameless world, obviously of human origin, were compiled (as an amplifier, as a probe) by an ancient, long-lived and slow former organic species with body parts

> "made of metal on a million different planets and their nerve impulses were light; this makes a big, big beast.... We think they made us for a joke or a trick, for they themselves couldn't go Inside ... we couldn't tell why they had made us or what they were going to do with us.... They are all dead now, of course."
> "Why?" said Jai, though of course he knew why.
> *We killed them,* said Evne. *What else*? [pp. 188–89].

34.
Lester del Rey,
Pstalemate (1971)

The American author Leonard Knapp (1915–93), who wrote under more than a handful of pseudonyms such as Philip St John and Erik van Lhin, was best known as Lester del Rey, especially after he and his fourth wife, Judy-Lynn, launched the Del Rey fantasy and science fiction imprint of Ballantine Books. Its first enormous hit was a Tolkien knockoff, *The Sword of Shannara* by Terry Brooks. Much of his own work was routine, but some made a mark, in particular his novella of humanity's war against God after the deity shifted alliances to an alien species, "For I Am a Jealous People" (1954), and this psience fictional investigation of the implications of precognition, *Pstalemate* (1971).

Harry Bronson is a 30-year-old New Yorker, engineer and inventor of a revolutionary motor that he and his associate are trying to market. His mother

Martha (whom he believes died twenty years ago in a bad car crash, along with his physician father) is incarcerated in an asylum for the insane, convinced that she is dead and being tormented for her sins. In fact she suffers the gift or rather curse of seeing the future, and like all other known precognitives has been driven recursively and paradoxically into madness by the increasingly terrifying fate of her own ultimate and dreadful insanity. Harry senses that he will fall victim to this same fate, and struggles desperately to find a way out of the horror awaiting him.

The first intimation is a woman's alarmed but spectral voice, his mad mother's telepathic call although he knows nothing of this yet. He hears it again as he climbs tenement stairs to a regular fantasy gathering, the Primates, calling his name: "Henry!" Del Rey's pulpish background is evident in the overwrought treatment:

> The voice from his nightmare screamed at him, jolting him and cutting into his mind with a blast of raw fear that froze his lungs and heart. He caught his breath and staggered back a step, tense with the same horror that had sometimes brought him awake in a cold sweat from his sleep. He had to get out of here! Up there, something foreign to all human feelings was waiting for him, something that must never...
>
> Then he mastered it. A trace of dread remained, along with his surprise at being caught by it while awake; the call usually came only in the middle of deep and otherwise dreamless sleep. He took a slow breath, rubbing sweat from his forehead with his sleeve [15–16].

At the Primates' smoky gathering, he is inveigled by license-revoked physician Philip Lawson into taking part in a light-hearted ESP card test. After a run through the standard Zener cards with (improbable) zero success, Lawson hypnotizes Harry to relieve the traumatic blockage closing off his psi abilities. A canasta double deck of playing cards is shuffled, and Harry calls them—all 108—correctly. The probability of this success is fantastically low. "'I want to go home,' he heard himself cry, and it was the sound of a small and lost child" (29). What follows shakes him thoroughly: Harry begins to capture information from other people's minds.

Ψ

Unusually for science fiction, even of the psience variety, del Rey attempts to discover the possible scientific basis of paranormal events. He does this by having a famous mathematical Primate, Dr. Byron "Bud" Coleman, FRS, toss out several ideas. This is clearly a tip of the hat to a very highly regarded theoretical physicist, Professor Sidney R. Coleman (1937–2007), an aficionado of sf, friend of del Rey's, and perhaps the source of this sketchy analysis. Bud draws on Shannon's key work in information theory in considering the information density needed for different kinds of purported psi.

When Harry comments that telepathy would require a signal bandwidth beyond the capacity of the brain, Coleman disputes that as a glib dismissal.

A signal akin to an AM radio, he notes, carries some 5000 bits of data per second and so needs a 5 kilohertz bandwidth. That would serve to convey sub-vocalized speech. However, for the brain's neural network to radiate a useful signal it would need signals in the gigahertz range, indeed waves different in kind from standard electromagnetism. "No telling what the velocity of propagation would be.... No reason why it should be limited to ... the speed of light" (p. 61). No further elaboration on this improbable claim. What about tuning-in on a particular mental transmission? "No problem.... Like FM radio, where a good receiver will pick up one signal and reject another on the same frequency if there's even one decibel difference" (p. 62).

And precognition and PK? Precognition *per se* does not exist, Coleman explains. It's all telepathy but, since its information spectrum differs from what we know, it might well propagate in time as well as space. "Sensitivity must be in picovolts" and the transmitter most like your neural receiver is your own brain in the future, you can be warned of future dangers or benefits by tuning in to your own future knowledge. Psychokinesis, however, is declared impossible; its energy would have to be metabolic, and the brain has no sweat glands to dispose of heat. "The idiot trying to drive a nail or lift himself by mind power would jolly well soon run a fatal cranial fever" (p. 63).

Harry reflects that he now has a theory to work with, although most of it is functionally useless. In terms of what is actually known about psi from research, that is a fair assessment of Bud Coleman's analysis. (Ironically, one major current model of psi—that of Dr. Edwin May, former director of the Star Gate scientific wing—is doubtful about the reality of PK, but asserts that *only* precognition exists.)

Philip Lawson the disdained quack, it eventuates, is really Philip Bronson, Harry's father, a second-generation telepath.

Ψ

Meanwhile, a newsman from the Primates meeting has written up Harry's fantastic card score, drawing the censorious attention of lawyer/ guardian, Charles Grimes, custodian of his trust fund. Stop this nonsense at once, or be cut off. That would wreck his chances of developing his advanced engine. Another of Grimes' wards, Ellen Palmero, has gone missing, and Harry is warned not to look for her. Ellen's father, once a stage mentalist, strangled her mother to death. Psi and perhaps even fake psi is looking like something to flee with all urgency. Harry, though, is led by a kind of psychic synchronicity to a diner where Ellen is passing herself off as a fortune-teller. She sends him away.

At a crux, after meeting a young psychic woman who believes she is doomed to death within days, he is struck by a revelation that "came with the total, absolute certainty that was a part of true precognition":

> It was the horror he had felt weakly behind the suffering and decision of the girl.... Now it looked as appalling reality, a threat against all the future.
> Madness came first. He was going to descend into raging lunacy, as his mother and father had done.... Now the dark and raging perversions of a mind turning against itself played through him...
> It was no longer his own self he felt, but an alien thing, an entity foreign to all he valued of himself! It was the ultimate evil—a demoniac possession, an alienness struggling to take him over, to dissolve his personality and shunt him aside, to use his body as a puppet.... Even now it seemed aware of him and was reaching through his telepathic channels for his mind [p. 74].

Harry succumbs to this atrocious presentiment, regressing into catatonia, and Ellen struggles to free him in scenes of surprisingly frank sexual discovery, begun when they knew each other as children. They marry quickly to avoid the social opprobrium of unwed coital bliss, such as it is, given Ellen's instilled belief that sex is filthy and degrading (rather reminiscent of the sanctimonious poison inflicted on other luckless women in a number of these psience fiction novels, notably Sturgeon's).

The explanation for this Alien Entity is simplicity itself: of course, Harry and other precognizers are foreseeing their potential future superhuman status. As they approach that condition, this can open a channel for future-self to reach back and aid past-self, even though doing so must necessarily require some amount of personality rewriting. But because of the implicit paradox in this re-entrant or retrocognitive process, past-selves are vulnerable to a horrified sense of their own personal extinction, such as a child might experience if she fully and empathically appreciated what kind of creature she would be at the age of thirty, let alone sixty or eighty. Perhaps for this reason all humans, as beings of self-aware consciousness, find it painfully difficult and aversive to imagine their final state—even those who ardently believe in a glorious and sacralized afterlife.

When finally Harry is linked to Ellen's mind as she goes insane, he understands that this violent swirling mass of filth is partly from her own warped childhood but much more a reliving of the brutal primitive lives of all her ancestors back to the split from earliest speciation. So, like his father the surgeon, he snips and tucks her brain/mind, clearing away these remnant horrors of the evolutionary process, leaving her for the moment like a child—reminiscent of Barbara D'Courtney in *The Demolished Man*. And his future self makes itself carefully available: "The future opened calmly to him. It came like a memory, with a partial sense of what had gone before" (p. 186). And in a moment that echoes Sturgeon and Clifford Simak, he recognizes an additional entity, not his future self, of which he has been dimly aware:

> There was a sense of opening outward—and the presence was there. Again it came with the feeling of immeasurable distance and gulfs beyond imagination. It came on its never-ending lonely search, reaching softly toward him. It was totally alien, yet warm with friendliness beyond that which humanity had learned [p. 187].

Again, then, we find this recurrent theme of psience fiction: life can be vicious and lonely, but the company of other embodied minds, even alien minds, can burst through the pstalemate of human isolation and find acceptance, friendship, even love.

35. Robert Silverberg, *Dying Inside* (1972)

More than a third of a century after its paperback publication, the notable critic and reviewer Dr. Michael Dirda offered a retrospective review of Robert Silverberg's finest novel, a history of the decline of a man's telepathic powers. Dirda asserted that

> It's insane that "Dying Inside" should be subtly dismissed as merely a genre classic. This is a superb novel about a common human sorrow, that great shock of middle age—the recognition that we are all dying inside and that all of us must face the eventual disappearance of the person we have been. More and more, as time goes by, our bodies break down, our minds start to lose their quickness, and, suddenly, inconceivably, our best work is behind us.[1]

While this is true, and the metaphorical aspect of preeminent concern to the litterateur, to a student of psience fiction the value of this closely observed naturalistic record of David Selig—forty-one, beyond the middle of the journey of his life, Jewish Ivy League graduate and failing telepath—is its unwincing portrait of psi in action and inaction. This does not mean that Silverberg's rendering is valid, based closely on real world psi reports, but his fiction has the value of lucid imagination: it clarifies what might be the case, if psi were anything like how he describes it, and what follows if it is. *Dying Inside* is one of perhaps the three richest novels of psience fiction. Silverberg's novel about a precognizer, *The Stochastic Man* (to which we shall return shortly), is almost in that class.

Silverberg has confirmed the necessary point that the novel is not, of course, really about telepathy but rather, as Dirda speculated, about the universal affliction of aging and loss of power and potency. He told me: "The telepathic stuff is just the narrative vehicle. I read some Rhine etc., but it certainly doesn't represent my own paranormal experiences, of which there have been none. The book is often mistaken for explicit autobiography. At some science-fiction convention years ago," he added with his distinctive dryness, "a woman who had just read it said, 'I didn't know you were one of us!' I replied, 'If I were one of you, you would have known it long before this.'"[2]

What's more, if cognitive psi works by conscripting and repurposing our ordinary senses and internal imagery, as seems plausible, an exceptionally vivid psychic ability might routinely interfere with the utility of those ordinary senses. Distraction by clairvoyant voyeurism while driving fast on a freeway is precisely the kind of risk Silverberg's tragic, defeated character David Selig runs even as he finds his craft ebbing. (Selig's psi has always been receptive only.) While in the real world powerful spontaneous evidence for the existence of psi has been found in the anecdotes of those who have evaded death due to a prophetic vision, it remains obvious that many people continue to die in accidents that might have been avoided with the help of psi. Conceivably, in some cases at least, the deleterious side-effects of psi, and its interference with necessary survival functions, might lead to the gradual attenuation and final shut-down of that faculty, a macro-scale version of what is known as the decline effect in parapsychology experiments. *Dying Inside* presents a scarifying image of a once high-functioning telepath:

> I haven't had much love in my life. That isn't intended as a grab for your pity, just as a simple statement of fact, objective and cool. The nature of my condition diminishes my capacity to love and be loved. A man in my circumstances, wide open to everyone's innermost thoughts, really isn't going to experience a great deal of love. He is poor at giving love because he doesn't much trust his fellow human beings: he knows too many of their dirty little secrets, and that kills his feelings for them. Unable to give, he cannot get. His soul, hardened by isolation and ungivingnesss, becomes inaccessible, and so it is not easy for others to love him. The loop closes upon itself and he is trapped within [p. 52].

Selig is not without love, at times, though. Yet it seems to become poisoned, inevitably, one way or another. He notes that everything was going well with Toni, his beloved, because he had "taken special care not to see too deeply into her" (p. 52), and he remains sure she loved him in return. This cautious and accommodating reticence comes unstuck when Toni tries her first LSD trip, Selig standing by to keep her company lest she run into trouble. Of course he gets the worst possible contact high, and rejects everything he sees in her naked soul. "You're trying to bum-trip me, aren't you, David? ... Why are you messing me up? It was a good trip" (p. 63).

Not for Selig; He decamps to the room of the drug-savvy gay couple down the hall for a tranquilizer, but by the time he has recovered Toni is gone, taking all her possessions. The way to remain sane as a telepath is to eschew close emotional contact with those for whom one has strong feelings. The loop closes upon itself; he is trapped within. After such knowledge, what forgiveness? Perhaps the only cure for the invasive, involuntary penetration of another's soul is the abandonment of psi. With this conclusion (whether valid in the real world or not), we move entirely outside allegory or parable of the quotidian, because aging and senility are not adaptations to an excess of power. That is what makes *Dying Inside* psience fiction rather than a coded conventionally realist novel about a man with receding hair, rheumy eyes and a sagging penis.

It turns out that Selig is not the only psi-afflicted human in New York, let alone the world, although because his access to other minds is limited to people physically near him he knows of only one, Tom Nyquist, whom he stumbles upon. Nyquist has met five others in the U.S., including Selig; they drifted apart. Nyquist refuses to feel guilt or shame in his selfish dependence on psi to locate available women and to make easy money. Not that Selig resiles from these opportunities (he was in the stock market for a time, but abandoned it, and finds plenty of women for sex), but his parentally inflicted sense of worthlessness does him in. He resents and hates his sister Judith: "Why do you hate me so much, Duv?" "I don't hate you. And we were about to get off the phone, I think" (p. 34). (It is perhaps necessary to mention, of such an apparently autobiographical work, that Silverberg was an only child). The threatening dissolution of his ambiguous gift promises a reconciliation between them, which would not be possible had she shared his affliction.

And yet this dreadful gift has brought him wonderful moments of mystic insights, immersion. As a boy, he visited the country and found a grim, forbidding German whose deep unconscious he visits. This is the very contrary of Dan Morgan's violent, terrifying id-unconscious. He enters old Georg Schiele's ocean depths and finds an ecstatic. The farmer

> stands in the rich soil of his fields, leaning on his hoe, feet firmly planted, communing with the universe. God floods his soul. He touches the unity of all things.... How can such a bleak, inaccessible man entertain such raptures in his depths? Feel his joy! Sensations drench him! Birdsong, sunlight, the scent of flowers and clods of upturned earth, the rustling of the sharp-bladed green cornstalks, the trickle of sweat down the reddened deep-channeled neck, the curve of the planet [p. 81].

This is part of what Selig will lose, when the power is abstracted away from his being. But with it will go, we only hope, the inauthenticity of a life lived only in reflections from the partial experiences of others. "Living, we fret," he tells us in his parting words. "Dying, we live…. Until I die again, hello, hello, hello, hello" (p. 245). However committed a parapsychologist might be to the reality and wonder of the paranormal, I think he or she might rest a moment to consider Duv Selig's liberation, his final soothing freedom from psi.

36.
Katherine MacLean, *Missing Man* (1975)

Born in 1925, Katherine MacLean is one of the indomitable women science fiction writers whose excellence in the 1950s and beyond disproves the canard that sf readers prior to about five years ago (or fifteen, or twenty five, or thirty five, depending on the age of the complainant) so vehemently disliked women's fantastika that editors refused to publish any. MacLean published notable stories in that hallmark magazine *Astounding* under Campbell's reign, starting in 1949 with "Defense Mechanism" and "Incommunicado" in 1950 (submitted three years earlier), with the razor-edged whimsy of "Pictures Don't Lie" (1951) and "The Snowball Effect" (1952) in *Galaxy*, and others.

Dismayingly, she became trapped in the Dianetics cult for a time, but her later fiction proved as impressive as the earlier work—most notably, her later sequence of three "Rescue Squad" stories for *Analog* (in 1968, 1970 and 1971—the third won a Nebula award) that became the basis of *Missing Man* (1975—nominated for a Nebula), a remarkable world-building novel set in 1999 after an apocalyptic crisis that destroyed much of New York, and forced a total rebuilding of its civic substructure. ESP forms a major element of the narrative (the German translation is *Der Esper und die Stadt—The Esper and the City*) but it is simply accepted by law enforcement and other instrumentalities; there is no need for proof of its existence or terror at its implications. In this sense, the book is a pure psience fictional distillate of Campbellian attitudes to paranormal abilities in the preceding decades.

George Sanford tells much of the story, which is interspersed by over-the-shoulder reports on the activities of others, sometimes merging into George's awareness and vice versa via his empathic/telepathic gift. He seems an amiable if meandering klutz for much of the opening, a former orphan and street gang kid who has just turned 20 and reveres his one-time gang leader Ahmed (Kosvakatis) the Arab, now a police operative dealing with lost or damaged citizens in trouble. Ahmed is not Arabic but Algerian French on his mother's side and Spanish Romany on his father's (p. 62).

This conglomerate ancestry is in some respect a marker of the cosmopolitanism of 1999 New York, after the battering that has rebuilt the city, while also partitioned into ethnic or tribal districts built inside walls from the ruins of the most damaged sectors. Police patrol the commons between them, but an understanding allows the Balkanized locales their bitter autonomy. The New York subways have been replaced by an expanded system of mobile chairs that attach to moving chains, a sort of version of Asimov's Caves of Steel and Heinlein's Roads that Roll. Several undersea cities house thousands of citizens, and come to horrific grief when criminal terrorists find ways to destroy them by turning their design features against them.

At the outset, once-chubby George has been without food for some weeks, and has managed to shed 80 pounds while building up serious muscle mass due to the free high-powered liquids he imbibes, dossing down in small urban communes somewhat reminiscent of San Francisco in the 1960s and Greenwich Village in the 1950s and '60s. He can't get a job because many menial tasks are computerized but more trenchantly due to his dyslectic inability to deal with the multitude of mandatory bureaucratic forms people keep insisting he fill out. Again this leads the reader to assume that George is mentally challenged, which is indeed the case but it turns out that the challenge is the profound but suppressed memory of a dreadful traumatic plane crash he caused, unintentionally and psionically, at the age of five. Combined with the cruel child rearing program devised by his programmer parents, this makes him unable to engage on any but a direct personal level with the officialdom of this rather libertarian yet bureaucratized state. And because he's so radically out of step, those officials can't make any sense of him—he's *refusing to fill out the forms!*—and only Ahmed's sponsorship allows him a hearing with the Rescue Squad.

It becomes evident that George is far from stupid. He just thinks differently, and luckily for him the free form floating nature of this future New York of 1999 allows him to get by without turning into a squalid gang rat, although he meets some of those eventually, led by a brilliant 15-year-old who has orchestrated the recent atrocities and plans more. It is never explained how Ahmed, a former gang leader, managed to get a degree by the age of 20 or so and a senior post with law enforcement by 21, or how the psychotic but learned Larry Rubsshov turned himself into a sort of Samuel R. Delany/Rimbaud/Foucault genius street kid poet cum terrorist. Larry's persuasive detestation of the permissive yet totally controlling technicians' social order is displayed in a scene where he and his gang hold George captive. He stresses that anyone can leave the city for a funded and housed utopia if they agree to be sterilized and never return. Many of George's street gang pals have already done this.

> They don't know how to be human beings. They like to read about being Tarzan or see old movies and imagine they are Humphrey Bogart and James Bond, but actually all they have the guts to do is read and study. They make money that

way, and they make more gadgets and they run computers that do all the thinking and take all the challenge and conquest out of life. And they give a pension to all the people who want to work with their hands or go out into the woods or surf instead of staying indoors pushing buttons, and they call the surfers and islanders and forest farmers Freeloaders and make sure they are sterilized and don't have children. That's genocide. They are killing off the real people [p. 108].

Worse still, these benevolent Skinnerian behaviorists purge criminals and undesirables of the special memories and urges that cause trouble, using a precise technique of neural destruction known as "mind wiping." It is yet another version of Bester's Demolition, applied by bored but self-satisfied experts on an industrial basis.

Larry adds an analysis of the growing impact of people he calls "autistic," those who love isolation and abstract thought. What is perhaps most notable about all this is how aptly MacLean, in the early 1970s, forecast some of the key characteristics of Silicon Valley Asperger spectrum culture thirty and forty years later.

Putting to one side his childhood trauma, is George an exemplar of this culture? In some ways, yet his psi gift is its exact contrary: his mirror neurons are working overtime, allowing his telepathic search beam to find and lock onto people in trouble, identify with them, communicate on a deep and sympathetic level that sometimes saves them from the soulless destruction their society is imposing on them. If it seems improbable that young people of fifteen or twenty could function this way, part of the explanation might be that the people of this future are actually very bright, if narrow. Here is a memory George retrieves from childhood:

> Every system becomes a system by excluding its opposite actions. In human nature all opposite impulses, repressed, do not fade. They accumulate and build up charge and fantasies. All old and lasting civilizations stabilized themselves by holding periodic ceremonies to release the charged opposite actions [p. 126].

In our world, this structuralist and poststructuralist dictum became grist to the academic mill a decade or two later, in the doctrine of the *carnivalesque*. It is attributed here to "the precise voice of the fifth-grade anthropology teacher." Perhaps it is not impossible that in MacLean's world this actually is part of the technicians' schooling for ten year olds.

In the Karmic Brotherhood Commune, George dresses for just such a carnival, and finds himself drawn to a display organized by sadistic fans of Aztec culture, whose inflictions upon their own kind are protected by the police. This scene and some others displaying crisis and George's clever yet almost unconscious response to danger and harm are bravura performances.

What is perhaps most remarkable about this novel by a writer brought to the world's notice in *Astounding* has been epitomized by the astute commentator James Nicholl:

> Even though the original novella (and the related stories) were bought by John W. Campbell, Jr. for *Analog*, and even though George fits a profile for a certain kind of Campbellian character, large swaths of this book read like pointed critiques of Campbell's hobbyhorses…. MacLean takes aim at ideas that range from removing the unfit from the gene-pool to the desirability of a psychic elite running the world (an idea that is raised, critiqued, mocked, and then discarded in the space of about five pages). and she is not kind to Campbellian enthusiasms.[1]

When finally George rebels against the status quo and is subjected to the inevitable mind wipe, his psi gift provides what is literally a mirroring that floods back into the quack "treating him," the technician at the power controls, and finally almost everyone in the city with that particular mental set. Using this control over others, he frees himself and, Gully Foyle–like, attains teleportation "involving a mixture of space, time, the fourth [spatial] dimension, and sheer imagination" (p. 217), vanishing in front of Ahmed from the hospital where he is recovering and reappearing "standing on the hospital bed."

Ahmed is shaken, blurting out quite uncharacteristically, "You ESP people are all freaks." It seems a fair estimate.

37. Robert Silverberg, *The Stochastic Man* (1975)

If Silverberg's *Dying Inside* was unusual (set in 1976, published just four years in advance) for a science fiction writer more accustomed to longer perspectives, often very much longer, his novel of precognition is a step only slightly forward to then-futurity. Serialized in the *Magazine of Fantasy & Science Fiction* in mid-1975, it is set around a rather gaudy 1999, on its way to the false millennium of the year 2000 (which was, strictly, the close of a millennium, not the opening of one). It charts a political race for the presidency of the United States with mid-30s narrator Lew Nichols opening the doors of the future for his chosen and fore-known candidate Paul Quinn. Nichols is guided, in turn, by a wizened financial speculator from the near-slums, Martin Carvajal, who has built up a fortune over the years and used none of it to aggrandize himself. He is precognitive, and begins by providing Lew with hints and finally by teaching him the secrets of secure prediction.

This is all wonderful fun until Lew foresees his candidate, the charismatic mayor of New York, blazing a path to a Hilteresque presidency and beyond. At the same time, his gorgeous wife with their open marriage slips away into the loony embrace of the Transit Creed, a sort of Dice Man lifestyle where free choice is abandoned in favor of randomness—the very principle of stochastics, or probability rather than strict causality, that Lew has previously used in his business.

This is a dire outcome in a psience fictional universe, and indeed in our own. It is one thing to profess a faith in divine predestination or retrocausation or the Block Universe of special relativity (where all times are equally real and "now" in their own frameworks). It is another to *know* that this is the case, that there is no escape, that your future was fixed before your birth and those with the gift of precognition can trace your destiny even if there is no deity pushing from behind, that you are set on rails in a purposeless universe.

Carvajal has developed a quietistic theory of this condition. He tells Lew:

> The ordinary person confined to a single universe can roam his memory at will, wandering around freely in his own past. But I have access to the memory of someone who's living in the opposite direction, which allows me to "remember" the future as well as the past.... If the two-time-lines theory is correct....
>
> The things you *see*—do they come to you in reverse chronological order? The future unrolling in a continuous scroll, that sort of thing? ...
>
> No more than your memories form a single continuous scroll. I get fitful glimpses, fragments of scenes, sometimes extended passages that have an apparent duration of ten or fifteen minutes or more, but always a random jumble, never any linear sequences.... I learned to find the larger pattern myself, to remember sequences and hook them together in a likely order [pp. 115–16].

Silverberg seems here to be preparing the ground for an utterly deterministic trap, but we hope, surely, for a possible escape route. Lew certainly does; he is revolted by Carvajal's doctrine of a cast-iron future, even as he uses the old man's patchy and knit-together forecasts to guide his candidate. And as in some Grimms' fairytales parable, the greater his success the more unease is spread by this unnerving accuracy. Candidate Paul Quinn is too smart and pragmatic to blame demons or sorcery, but he cannot trust a guide whose methods are unintelligible. Lew is fired. His wife is gone. His life, as he begins to experience it also in advance, is dismaying even as his non-stochastic skills broaden and deepen.

Since we are here concerned above all with the themes and mechanics of various kinds of psience fiction, it is worth looking closely at the ingenious, even *tour de force* designs of this terrible trap.

Ψ

In his first months and years as Quinn's advisor, Lew tests the waters of the future by making numerous phone calls to Carvajal, filling him in on the

The Stochastic Man (1975) 157

momentous tipping point events of the day. A time loop is formed. These data-dumps cannot help him now, but the information he provides in this regular way will enter his guru's consciousness and hence be available for messaging back to the past—to Lew's present. He comments (and one detects Silverberg winking at us):

> Over the remaining year of Carvajal's life he would become a unique repository of future political events. (In fact he already was that repository, but now I had to follow through by making certain he received the information that we both knew he was going to receive. There are paradoxes inherent in all this but I prefer not to examine them too closely.) [p. 149].

Parapsychologists (and, increasingly, physicists) have necessarily spent time and energy trying to resolve the apparent paradoxes of altering the past by using foreknowledge of future consequences. The general solution appears to take this form: the future can influence its past if the new outcome can be attributed to a chance error in the experimental preparation, or an unpredictable quantum state collapse. Another way of looking at this is to regard only *previously unobserved states* to be suitable objects for backward interference. You can't make an outcome *unhappen*, once it is a matter of record (including growth rings in ancient fossils, light arriving from a distant supernova, etc.). But you can *contribute* to its causal chain. Experiments have suggests, for example, that practicing the answers to a quiz *after* you have answered it can increase the number of correct answers you mark.[1] In many instances like this, time's causal arrow is reversible and additive.

A rather different model calculated by Oxford quantum physicist David Deutsch suggests that time travel can happen without paradoxes (even if you kill your parents before you are conceived) by automatically opening one or more extra histories of the universe, allowing you to continue in a slightly different sub-universe. This requires a Multiverse, of course, in which, it is conjectured, all possible states of all particle interactions really do occur, or can be made to occur.

This last version is more or less what Carvajal accepts as the way things are, although he also feels that most of the alternative futures revealed in his kinds of vision are bogus, shadow realities that merge or fade away into the one true destiny. He has seen his own death for many years, and has become inured to its horror. What will be, will be. Lew is less convinced when he begins to see different deaths for himself. The basic death is far off, but alternatives, some due much sooner, also assault him:

> A baffling torrent of images roared through my mind. I saw myself old and frayed, coughing in a hospital bed with a shining spidery lattice of medical machinery all about me; I saw myself swimming in a clear mountain pool; I saw myself battered and heaved by surf on some angry tropical shore.... Death,

again and again, coming to me in so many forms.... What am I to believe? I am dizzied with an overload of data; I stumble about in a schizophrenic fever, *seeing* more than I can comprehend, integrating nothing [pp. 205, 220].

Very well—but if this is so, how could Carvajal have become so immensely wealthy, in a future so labile, so underdetermined? He explains again, and to convey Silverberg's cunning here I must quote at length:

> Carvajal smiled and held up one hand, palm toward me. "Believe," he urged in the weary, mocking tone of some old Mexican priest advising a troubled boy to have faith in the goodness of the angels and the charity of the Virgin. "Have no doubts. Believe."
> "I can't. There are too many contradictions." I shook my head fiercely. "It isn't just the Quinn visions. I've been *seeing* my own death, too."
> ..."There are many levels of reality, Lew."
> "They can't all be real. That violates everything you've told me about one fixed and unalterable future."
> "There's one future that *must* occur," Carvajal said. "There are many that do not. In the early stages of the *seeing* experience the mind is unfocused, and reality becomes contaminated with hallucination, and the spirit is bombarded with extraneous data."
> "But—"
> "Perhaps there are many time lines," Carvajal said. "One true one, and many potential ones, abortive lines, lines that have their existence only in the gray borderlands of probability. Sometimes information from those time lines crowds in on one if one's mind is open enough, if it is vulnerable enough" [pp. 223–24].

Is this truly Lew's fate? Will Paul Quinn prove to be a blood-spilling psychotic, an American Hitler? We readers never find out. Perhaps all the possible Lew Nichols learn all the possible answers, each of them in an separate world. That would truly be a stochastic outcome, if the probabilities are imagined as equally valid, equally real, superposed, all contributing (somehow, through quantum magic) to the one world that is not a phantom. That is, in fact, our jointly experienced world, as represented here by Silverberg's psience fictional vision.

38.
Octavia Butler, *Mind of My Mind* (1976)

Most discussions of Octavia Butler's writing career begin by mentioning that she was one of the few black women science fiction novelists, that

Mind of My Mind (1976)   159

she died tragically young (at 58—she was born in 1947, died in 2006), perhaps after suffering a stroke outside her home (she was troubled by serious hypertension) and injuring her head as she fell. She had been granted a MacArthur "Genius" Fellowship as well as winning two Hugos and two Nebulas. Her major psience fiction sequence is known as the Patternist series—*Patternmaster* (1976), *Mind of My Mind* (1977), *Survivor* (1978), *Wild Seed* (1980), and *Clay's Ark* (1984). These do not proceed in chronological order, and the best entry point is probably the second, although the fourth visits earlier millennia and centuries of the Immortal male character, Doro, who can shift from one body to another at will, leaving a trail of corpses behind him, and the long-lived woman shape-shifter Anyanwu, known later as Emma.

Doro is a sort of eugenics gene-engineer, originally a Nubian in Egypt four thousand years ago, now roaming in bodies of any ethnic group as suits him. He has impregnated thousands of women through the millennia, then done the same with their daughters, arranging marriages between offspring who shown paranormal abilities, keeping them as psychic slaves within African and later New World "seed villages." Anyanwu becomes his wife, bearing many children—a large proportion of whom prove to be damaged, and expendable—and to Anyanwu he weds his favorite son, psychokinetic Isaac, in hopes of creating an even more marvelous child. It is a disturbing parable, blending unexpurgated power, violence and sexual fantasies from folklore and myth with a master-slave iconography that was enacted in the real world upon the bodies and minds of many Africans, as Butler was bitterly aware.

Given this quite consciously parabolic aspect (and Butler's later work would contain novels explicitly titled *Parable of the Sower,* and *Parable of the Talents*), is it appropriate, or even an appropriation, to dub the Patternist sequence "psience fiction"? Yes, clearly it is appropriate. The centrality of paranormal powers such as PK, telepathy, healing and other uncanny abilities, as well as non-canonical capacities such as shape-shifting (Anyanwu can modify her body into the cellular arrangements of land animals and dolphins, in an extreme extension of the "womanly talent" in Anne McCaffrey's psi novels), make these tales prime exemplars of the sub-genre.

In *Mind of My Mind*, Anyanwu has been known as Emma for a century and a half, mostly presented to the world as a crone. Her grandchild several times removed, Mary Larkin, is a Latent telepath approaching Transition into the Active condition. This is the cruelly painful harrowing of the hell that is other people, their worst terrors flooding her unprotected mind, pushing her toward either madness and suicide or maturity, protected behind a shield. During this horrifying apprenticeship, she is married to telepathy-gifted Karl, at Doro's command, although he is already in a relationship with a non-psi, Vivian. We experience Mary's transition into psychic maturity, in all its frightened secondhand pain, and the emergence of a new kind of psi-bonding with Karl and five distant strangers, the first of the Pattern groups:

> Their thoughts told me what they were, but I became aware of them—"saw" them—as bright points of light, like stars. They formed a shifting pattern of light and color. I had brought them together somehow. Now I was holding them together—and they didn't want to be held.
> They tore at me desperately...
> They rested grouped around me, relaxed...
> I realized that there was something really proprietary about my feelings toward them [p. 61].

The psychic condition is emotionally unstable, Butler informs us, and has many hazards. Karl's mother was a drunk *latent*, "too sensitive," like Mary's own, Rina, a whore. When Karl was a child she held his hand over the flame of the gas range, holding it there as it charred. "Doro came to see her later that day. I wasn't even aware of when he killed her.... Doro's healer arrived ... one of Emma's granddaughters. Over a period of weeks, she regenerated the stump" (p. 49). Another latent in repeated pain and confusion is Clay Dana, at 30 a year older than his devoted brother Seth. Planning to live at the edge of nowhere, miles from the mental noise of a small town, they receive the overwhelming message *Come*, drawing them to the west, toward Forsyth, a Los Angeles suburb, home of Rina and Emma. So do Rachel Davidson, a healer, Jesse Bernarr, Ada Dragan. All these various mutants seem to be trapped in an immense and fluctuating internet of the mind, a mental landscape we suspect Mary and her Patternists are destined to rule.

Butler shows us others in this group of flawed active psi mutants, and the one latent who finally opens into his full power, all drawn against their will to Forsyth where, Doro warns Mary, they will all detest each other to the point of murder. There is a lot of barely throttled fury in Butler's imagined worlds. In a notable psience fictional scene where two actives attempt to whip up an attempt to kill Mary, whose pattern holds them all in thrall, this young woman, barely through transition herself, sucks energy in a hungry spasm from the ring leaders, Jesse and Rachel, without murdering them:

> Abruptly, he dove at me through his strand of the pattern. I had no warning. He acted on impulse, without thinking.... His strand of the pattern struck at me snakelike. Fast. Blindingly fast.
> I didn't have time to think about reacting. What happened, happened automatically...

Mind of My Mind (1976) 161

He was mine. His strength was mine. His body was worthless to me, but the force that animated it was literally my ambrosia—power, sustenance, life itself...
Rachel ... thought Jess was dead. She, a healer, thought he was dead, but I knew he was alive.
She finished turning. She was going to rupture a good sized blood vessel in my brain.
I took her [p. 133].

It is a powerful moment that underlines the implicit equation of Doro, Mary and other powerful actives as vampires—not suckers of blood, but of "life force." This need is, indeed, the primary motivation for the seed villages and Mary's Patternist wheels-within-wheels: to provide ample feedstocks for those at the tip of the pyramid. It is a social model closer to feudalism or Nazism than to outright slavery or corporate capitalism, but it is clearly not the kind of gestalt advocated by some of the psi fiction of Sturgeon, for example.

When Mary is finally ordered by her father to stop recruiting new latents, saving them from the torments of their condition (which is rather like hearing screeching out-of-tune radio broadcasts with no surcease), she refuses, while knowing that rebellion will cause Doro to kill her. She avoids this fate in the last pages of the novel by drawing on the dispersed energies of her Pattern, sucks Doro into the maelstrom, and performs on him the vampiric feast he has intended for her. Doro dies, along with ten percent of her 1500 Pattern incumbents. Her ancestor Emma dies as well, perhaps in hopeless despair. The future is thus open to a world interpenetrated by Patterned collectives—and indeed in two volumes beyond this one, an interstellar destiny unfolds, perhaps equally noxious by the standard of Enlightenment ethics.

The far-future Patternists in *Patternmaster*, Butler's first novel, are telepathic and gifted in other psionic skills, highly intelligent, and not dissimilar to the post–Apocalypse quasi-Black masters of Robert A. Heinlein's *Farnham's Freehold* (1964). That Heinlein novel has been generally condemned as racist although it made the same kind of sociobiological point marking the Patternist sequence: that people, human or posthuman, adopt local ethical systems, introjected by the subservient, that suit their leaders.

And perhaps the key virtue in Butler's sequence (although it is foreshadowed elsewhere, especially in series by Marion Zimmer Bradley and Anne McCaffrey) was pointed clearly by critic T.M. Wagner:

Many SF and fantasy novels have depicted the prospect of being superhuman—possessing psychic, telepathic, and other extraordinary powers—as eminently desirable, and comic books have practically made this notion their *raison d'être*, elevating it to the ultimate heroic archetype. Octavia Butler presents quite the opposite view in *Mind of My Mind*, an often gut-wrenching tale in which becoming superhuman is the first step to losing your humanity.[1]

39.
Joan D. Vinge,
Psion (1982)

On the planet Ardattee where the underclass live in squalor and the wealthy reside gracefully in Quarro at the top of immense towers, Cat is a street-wise seventeen-year-old urchin in Oldcity. He has vertically slitted pupils, hence his name, and without knowing it he's the offspring of a human and an alien Hydran from the Beta Hydrae system. (On the face of it, this is about as likely as a Rhododendron mating with a moo cow and producing a green calf with deciduous eyelashes.) He is a telepath, also without knowing it due to a frightful shock in childhood that has blocked his psi ability.

Captured by a Contract Labor press gang, he undergoes rather harsh therapy at the Sakaffe Research Institute that unlocks this Gift, making him a detested psion in a squad of others who can mind-read, teleport, PK, etc. He is guided by Seleusid Interstellar corporate telepath Derezady Cortelyou and Jule taMing, a teleporting empath who teaches him to read and write. (Her wealthy and powerful family controls Centauri Transport, so being a psion has led to her banishment here). Dr. Ardan Siebeling, a telekinetic some two decades older than Cat who runs this group, is Jule's lover and inexplicably hates Cat.

Derezady explains psi to the youth:

> the mind was a net of electric fire—nerve fibers reacting to every sensation and image, every thought and feeling that let human beings interact with life. In most humans the input and reactions were woven into a snarl that even biofeedback training could barely begin to unravel. Psions were born with something more—a set of self-controls that let them weave the snarl into patterns and, more than that, to tap and use a kind of energy normal humans were blind to [*Alien Blood*, p. 27].

The incomplete series to date (but presumably not to be continued) is *Psion* (1982), *Catspaw* (1988) and *Dreamfall* (2004), set in a highly detailed world-building background with equally detailed characters and often families. At the end of *Alien Blood*, which combines the first two novels, Vinge wrote: "Cat is a character I began to write about when I was seventeen [that is, in 1965]. I plan to write a series of novels about him and what happens to him though his life" (p. 634). This long-term plan was interrupted by a serious road accident in 2002 that left her brain-damaged; she finally returned to writing in 2007, but not of any note. The tenor of these novels, by curious advance resonance with Vinge's medical problems, is notably grim, with Cat's abilities silenced several times by traumatic events. One of these is a necessary killing, saving his own life and that of several others but leaving his Hydran-inflected mind once again in blocked retreat from telepathy. In this, it is somewhat reminiscent of the fate of Victor Coleman in Dan Morgan's *The Country of the Mind* (1975).

The Hydrans, it turns out, have evolved an extreme aversion to violence, which makes it difficult for them to resist the arrival of humans in a colonizing surge through the galaxy once superluminal flight is invented. Navigating a path faster than light requires a sort of quantum computer (not yet under development in the real world when the first volume was published) using the very rare element 170, Telhassium. "They lock data into the electron shells, and they can run a whole planet's information system on a crystal as big as your thumb.... The big cargo ships only carry a little more telhassium than an entire planet uses" (p. 29). Rubiy, a psion criminal with prodigious and multiple gifts, calling himself "Quicksilver" (a regrettably comic book choice by Vinge), plans to seize command of the element. Only a suitable "joining" by a team of psions working for the corporations and the Federation Transport Authority can defeat him. A joining is like a Sturgeon gestalt: "It's a meeting of minds, between a telepath and another psion, so complete and unguarded that their minds become one—each open to the other totally, with nothing held back. Their psi powers are heightened ... they do things they could never do alone" (p. 33).

Cat's gift, brought forth and trained, might be especially powerful, capable perhaps of besting Rubiy. Trained telepaths, Derezady noted, "could sort out the strands of image that patterned someone else's thoughts ... locate one particular pattern, follow it along all its branching ways to their scattered ends and back" (p. 34). This image of the mind as a sort of spiderweb, or even the World Wide Web that did not exist when Vinge was writing the first volume, is presumably functional rather than literal in a neurological sense, as it was in Dan Morgan's version of mind and brains and psi.

But the logic and penetration of the telepaths has some annoying limitations. In his first meeting with Ardan Siebeling, Cat pocketed "a clear ball that looked like crystal ... with a gold and green insect caught in flight inside it" (p. 60). It creates lovely memory pictures in his mind. Siebeling recovers it, dismisses him from the group as a thief. "It shouldn't respond to you, you couldn't even understand." Minutes earlier, Cat and Jule were reflecting that Ardan was a man "bleeding inside, there's something so sad, so terribly deep" (p. 58). Yes, he'd "had a Hydran wife, who had a half-breed son somewhere in the galaxy" (p. 60). It should not require a telepath to notice that Cat must be Ardan's lost son, yet this glaringly apparent insight is locked away from each of them.

The novel and its sequels continue as a picaresque, Cat working for a time for the dread Quicksilver, cast into involuntary servitude in the mines of the planet where his Hydran kinfolk are subsiding into extinction, at last confronting Rubiy and killing him to save not only himself but his father:

> "What did your kid look like?"
> "My kid...." He stopped, and took a deep breath. "He probably would have looked a lot like you, actually."
> "I wish I could have met him."
> He glanced at me and smiled just a little.

No doubt.

Psion ends in a psience fictional contest between the psi criminal superman and a joining of Cat, Jule and her beloved Ardan. Weapon in hand, Cat is tempted by Rubiy. He caresses Jule's back, and Quicksilver's lure is open and plausible: "Take her, if you want her? ... Use her, stop being a fool. Don't you know who she is; what the taMings stand for? They're one of the most powerful and corrupt shipping families in the Federation ... they make the systems work, our suffering, our despair.... Hate them" (p. 213).

The weapon in Cat's hand is pulled toward Siebeling, and fired by Rubiy. Siebeling is burned, and apparently dead. The gun moves on to Jule. The joining occurs. "The gun jerked in my hand ... [Rubiy] died. And his death was my own, in agony that exploded like a star and vaporized my soul inside shattering bonds of light" (p. 218). His Hydran heritage, written in his DNA, again obliterates his psi ability. "I'd never be a telepath again.... I'd killed and I'd survived it, and that was my punishment: to come full circle, to be a walking dead man, blind and alone and going nowhere" (p. 219).

It is a poignant, rewardingly downbeat conclusion. But actually it isn't the conclusion, just a pause, before the second and third volumes, where Cat learns a great deal more about the wickedness of the powerful, and grows up. Perhaps one day Ms. Vinge will recover fully her imaginative powers and continue, or allow one of her ardent fans to complete, the story of Cat the anguished psion.

40.
Lucius Shepard,
Life During Wartime (1987)

Catapulting to prize-winning prominence in the early 1980s, Lucius Shepard (1943–2014) effortlessly surpassed the achievements of the previous generation of once-routine sf writers—Robert Silverberg, Harlan Ellison—who had remade themselves more ambitiously in the 1960s and 1970s. But Shepard was already in the middle of his life. He had experience of the Vietnam war (perhaps as a reporter rather than warrior), had traveled widely in dangerous places from Latin America to Afghanistan, performed in rock bands, married and divorced.

His work is most affiliated with the baroque lushness of the magic realists of Colombia and Argentina: Gabriel García Márquez, Julio Cortázar. The superb opening of *Life During Wartime* first appeared in *Asimov's* sf magazine as the novella "R & R" and won 1987's Nebula and Locus Reader's Awards, but the novel, published without conspicuous genre trappings, was hailed as well by non-genre reviewers and critics. Amidst its other virtues, this magical realist novel is clearly psience fiction.

While Shepard's future did not quite come to pass, like almost all sf, its vividness and edge still bite thirty years later. Tel Aviv has not been nuked, Guatemala has not been invested by U.S. troops on the scale of Vietnam—but parallels are evident in Iraq and Afghanistan and the relentless drug wars in Mexico and other nations south of the U.S. border.

David Mingolla is a 19-year-old conscript, stationed in the preposterous Ant Farm, a military termitarium of tunnels and weapons quickly reduced to uselessness by guerrillas. It is a *reductio ad absurdum* of the protracted engagement in Vietnam, and serves as well, metaphorically, for the massively armored western military enclaves in Iraq and other oil-rich nations. What makes New York art student Mingolla special is his possession of psychic abilities. He cannot deny his own nature, and the impulses drawing him.

He meets Debora Cifuentes, a sexually compelling enemy psychic in her early twenties, running a game of chance: "her red dress and cryptic words; the runelike shadow of the wire cage—all this seemed magical, an image out of an occult dream" (p. 23). As indeed it is, or a foreshadowing. She tells him, "I have the gift myself, and I'm usually uncomfortable around anyone else who has it.... What do you call it? ESP? ... I'm surprised you're not with Psicorp" (p. 25).

In fact, he has been headhunted already by Psicorps, a merciless psi training branch of the armed forces distantly akin to, but far more ferocious than, the real Star Gate remote viewing program that was still deeply classified when Shepard wrote his fiction. Mingolla rejected the call, and tells Debora: "These guys think they're mental wizards or something, but all they do is predict stuff, and they're wrong half the time. And I was scared of the drugs, too. I heard they had bad effects" (p. 26).

But soon he is in the appalling "therapeutic" custody of Dr. Izaguirre, doped with a rare flowered weed that elicits and enhances psi powers—not just gappy precognition but an uncanny ability to shape and control the emotions of others. And the drugs do have very bad effects. They turn Mingolla, step by crazed step, into a sort of monster. It seemed no accident that his name is a near-homonym for Mengele, the Nazi experimentalist doctor from Auschwitz-Birkenau concentration camp who fled to Argentina and Paraguay. (Shepard denied conscious intent in this wordplay, stating that he drew the name Mingolla from a newspaper. But the story he published immediately after "R & R" was titled "Mengele," and the unconscious is tricky.)

In the 1950s, this plot could have been a standard *Astounding* psionics adventure seeded directly by John W. Campbell. In the hands of a luminous, tough-minded, politically embittered writer like Shepard, it moved far beyond those catchpenny limits. Sometimes the diction is a little too ponderous, portentous—"The memories of the dead men in his wake were weights bracketed to his heart, holding him in place"—but mostly Shepard's writing is various and fluently fitted to its purpose. It is raspingly obscene and brutal when the grunts talk macho trash, driven through superstitious rituals of self protection from magical assault, or sexually frank and fervent, or richly descriptive, drifting from lyricism to terror as one of the soldiers undergoes a psi-mediated attack by ... *butterflies*:

> A couple of dozen butterflies were preening on Coffee's scalp—a bizarre animate wig—and others clung to his beard; a great cloud of them was circling low above his head like a whirlpool galaxy of cut flowers.... Butterflies poured down the tunnel to thicken it further, and [Coffee] slumped ... the mound growing with the disconnected swiftness of time-lapse photography, until it had become a multicolored pyramid towering 30 feet above, like a temple buried beneath a million lovely flowers [p. 181].

Mingolla's progression is a descent through a nightmarish Purgatorio or harrowing of Hell: hallucinatory, compulsive, stripping flesh from bone. At the outset, he regards "the core problems of the Central American peoples," like his own, as being "trapped between the poles of magic and reason, their lives governed by the politics of the ultrareal, their spirits ruled by myths and legends, with the rectangular, computerized bulk of North America above and the conch-shell-shaped mystery of South America below" (p. 69).

Granted, this is presented as a patronizing appraisal by an untested youth, yet it eerily foreshadows the novel's psi-powered war between two old Spanish families that proves to be the hidden core of the mad, arbitrary conflict tearing at the Americas. Mingolla sinks ever deeper into this nightmare of myth, drugs, paranormal intuition, and personal command. He is held from the pit of absolute power by his bond with Debora, yet ruinously energized by it.

Is his final temptation to drag them together into a Faustian hell of survivalist banality, exemplified in the Talking Heads 1980s' song "Life During Wartime" put it: "Transmit the message, to the receiver/ hope for an answer some day." It seems so.

41.
Carrie Vaughn,
After the Golden Age (2011) and
Dreams of the Golden Age (2013)

While this study does not examine any of the vast number of comic book variants of psi and other paranormal powers, it is illuminating to consider Carrie Vaughn's reverse engineering from comics back to psience fiction as literature. *After the Golden Age* (2011)—and its one generation later sequel *Dreams of the Golden Age* (2013)—is a tribute to classic and recent superhero comics/comix, but also a solid attempt to rework those always-already recycled tropes into fresh literary form. If the first volume is more buoyantly enjoyable than the second, as well as poignant where loss and sorrow are appropriate, they combine as an intriguing speculative diptych on the theme of gaining and losing paranormal powers in a world quite like our own.

And *After the Golden Age* opens, true to its title, quite boldly *in medias res*. Celia West is not a superhero, but the forensic accountant daughter of middle-aged Spark (Suzanne West) and Captain Olympus (Warren West), top dogs of the vigilante Olympiad that protects Commerce City. They wear gaudy costumes, when they are doing a job; she does not, because she lacks any sign of a superpower of her own. If anything, when growing up she was overprotected by her psi-powered parents, but felt ignored by her angry father. In the sequel, Celia has matured, is fabulously wealthy and married to Dr. Arthur Mentis[1]—the world's greatest telepath—and mother of 17-year-old schoolgirl Anna, who does have a superpower (she knows where people are) but keeps quiet about that, and Bethy, who has none. Apparently the genes for paranormal powers are recessive.

These books are, strictly speaking, psience fiction only as a courtesy. The powers of the Olympiad and some of their children and grandchildren are specific and generally unique to each superhero, and there seems no taxonomic rationale to them. If one can whip up a storm at sea, and deluge nearby landscapes, this is done by wishing it to happen. No magical incantation is necessary, and no paraphysics. Stuff just happens, although there are limits to these exertions, and they can take it out of you. While many parapsychologists express their credence toward some claims of levitation, for example those of Daniel Dunglas Home and St. Joseph of Cupertino, few would expect superhuman speeds from transvection (the technical term for lifting off the ground and wafting away).

While telepathy is a central item in the list of claimed psi abilities, it rarely seems to take the usual psience fictional form of "hearing" the other person's voice "in your head." It is usually a matter of having a kind of uncanny sense that something critical is happening to a loved one, or maybe someone you've never met who is trapped in a crashing plane. Nor do we often hear reports of this kind of biological epiphany with the dawn of a psi capacity:

> [Anna had] been fourteen years old when her power awakened. The books and biographies about superhumans and their powers said they often manifested at puberty. It had for her grandmother and father. But she woke up one morning and her brain ached. Aspirin didn't help. It was like her entire mind cramped—she'd had her first period the year before, and this felt like that, only in her head instead of her gut. Then she seemed to *fill up*. Her mind expanded, taking on an extra sense.... Her awakening power was probably mental. Was she developing telepathy? Telekinesis? Clairvoyance? [*Dreams*, p. 50].

If superpowers are linked to inheritance, where did they originate? In Vaughan's novels, they are an accidental byproduct of a machine developed in the early years of West Corp, Captain Olympiad's wildly profitable corporation. Celia finds a document on file titled "Use of Directed Radiation to Induce Neurophysiological Responses, with the Intent of Encouraging Specified Emotional traits in Human Subjects" (*After*, p. 281). A psychologist, Sito, wanted to tame the sociopathic drives of habitual criminals. His method functioned by inducing loyalty—to community, in this case. "Why do you think the superhumans have all become crime fighters and not circus freaks?" Celia asks. "Something inside them drives them to it.... They're loyal to this city over everything else in their lives" (p. 283).

Perhaps this explains Superman's fondness to Metropolis, too, and Batman's to Gotham City, and Green Arrow's to Star City. It does not seem particularly relevant to psi exponents or researchers, who are on the whole a strikingly cosmopolitan group.

In short, the psi aspects of these two novels are no more than handy labels for the gifts traditionally attributed to the pagan gods, demigods and demons, especially the Greek and Roman pantheon much admired by educators in the 19th and early 20th centuries: Jove/Jupiter, with heavenly thunderbolts, Hermes/Mercury, with flight, Poseidon/Neptune, power over the seas, Demeter/Ceres, in charge of harvest fertility, Ares/Mars, commander of warfare, and so on.

These deities usually had the ability to read human minds, and share information depending on hierarchy; they could often achieve advance knowledge of the future, and dispense it via their sibyls, although often in ambiguous terms. They could see at a distance anything they wished to observe, so they were natural (or supernatural) clairvoyants. Does that suggest that human paranormal gifts of these kinds gave rise to imagined gods with clarified and enhanced powers of just this sort (and not, say, the ability to do partial differential equations in the blink of an eye, except in the case of Ramanujan who was guided by the Indian goddess Mahalakshmi of Namakkal)?

Or are many of the alleged psi powers nothing more than the wishful embodiment of our desires to exert our will over others, attract them sexually, harm our enemies, fly without a rocket booster, foresee the consequences of our actions (and win the lottery)? In any event, the goddesses did not, to the best of my knowledge, experience their awakening to power as similar to the onset of menstruation. That Vaughan makes this analogy is an acknowledgment that at least half the humans with explicit psi powers are, after all, female—and many of the most consequential, sibyls, shamans and mediums alike, were almost all women

> A Chukchee proverb declares, "Woman is by nature a shaman." Yet the female dimension of this realm of spiritual experience has often been slighted. Mircea Eliade believed that women shamans represented a degeneration of an originally masculine profession, yet was hard put to explain why so many male shamans customarily dressed in women's clothing and assumed other female-gendered behaviors. Nor does the masculine-default theory account for widespread traditions, from Buryat Mongolia to the Bwiti religion in Gabon, that the first shaman was a woman.
>
> In fact, women have been at the forefront of this field worldwide, and in some cultures, they predominate.[2]

For all their divergence from the reports of actually parapsychological phenomena, these novels are charming, entertaining, and thought provoking.

42.
Connie Willis, *Crosstalk* (2016)

Connie Willis (1945–) is one of the best-loved and widely admired writers of science fiction and fantasy, often blending emotionally gripping themes and relationships with hilarious or confusing madcap, helter-skelter, screwball comic turns. She has won more science fiction awards than any other writer, with seven Nebulas and eleven Hugos, for both novels and quite diverse shorter works.

Briddey (Bridget) Flannigan, a beautiful and stylish redhead, is a minor executive at Commspan, a communications company locked in a race to beat Apple's imminent release of a new smartphone. The book's second paragraph makes mention of *Gossip Girl*, which can stand in for the dynamic of not only the ceaselessly texting, tweeting, Facebooking and whispering workplace but also Briddey's quasi-Irish family, from great-Aunt Oona to the endlessly pestering older sisters Kathleen and frantic "helicopter mother" Mary Clare and her nine-year-old Maeve, who gets the book's final and telepathic line. Oona, who has never been to the "auld sod," insists on speaking in a dreadful fake brogue, and presses Briddey to find true love among the few male members of the Daughters of Ireland Gaelic poetry readings.

But Briddey is involved with her handsome and stylish boss Trent Worth, who insists that they undergo a minor brain implant linking the couple in a kind of enhanced emotional bond that conveys no messages other than a profound gush of mutual love. It is never explained what happens to people with such EEDs (the acronym is not spelled out, typically) if the love goes away. Nobody worries about the possibility of a future life filled with communicated boredom, irritation and bouts of rage.

As it happens, Briddey wakes from surgery with a strange voice in her head—not Trent's, not yet, but the intrusive presence of the odd genius with terrible clothes and hair who hides in the firm's basement inventing most of the devices they live by. This is C.B. Schwartz, regarded by Briddey's coworkers as a creepy "Hunchback of Notre Dame" although his redoubt is underground, blocking phone signals. C.B. had urged Briddey not to have the surgery, which can have horrid side effects. Now he's popping up in her head, and she cannot escape his surveillance. Luckily, as it turns out.

Schwartz, whom everyone assumes is Jewish, has been exploring the roots of his telepathic prowess for years. He explains that Briddey's (and his) Irish ancestry is responsible. "It has *everything* to do with it," he tells her, "—or more particularly, with the haploidgroup gene R1b-L21 the Irish carry."

It is worth noting that the correct scientific term is *haplogroup*, which denotes a heritable mutation passed along from a single ancestor in the DNA of a particular clan.[1] And while it is true that this "Celtic" R1b-L21 is the most frequent haplogroup in Ireland, it is also common in England and found in Spain, Norway, and France (which might explain Joan of Arc, who heard voices and gets a lot of coverage in the novel). Of course, we know that this genomic reference, with or without misspelling, is a psience fiction device, a piece of plausible hand waving with very little chance of being true in the real world.

The error in naming the genetic variant suggests that Willis was either oddly careless in her nomenclature, or she was deliberately introducing an error to flag the fictional nature of her explanation. Since there were complaints about a number of small errors in her massively researched diptych *Blackout/All Clear,* a World War II time travel saga, it's possible that this was just an inadvertent or copyeditor's typo—like her misspelling *Finnegan's Wake* (p. 479). This is a peculiar hazard in science fiction, which is grounded in signifiers with no real-world referent. Many sf readers mistakenly take it for granted, of course, that terms such as "ESP," "psi," "precognition," etc., are just such signifiers of allegedly "absent signifieds."[2]

Although Briddey starts receiving messages from other minds, C.B. attempts to convince her, by unsupported assertion, that psi is nonsense, either superstition or fraud. He repeatedly calumniates Dr. J.B. Rhine, whose research, he claims, was discredited:

> He counted almost everything as a correct answer, and the Zener cards were so thin, you could see right through the backs of them…. He clearly cherry-picked the data. He claimed that his subjects' telepathic ability took time to warm up and then faded when they got tired, which means he only used the periods when they had runs of correct answers…. Rhine messed with the data [pp. 149, 155].

Granted, in very early Zener cards (square, circle, wavy lines, etc.), a printing flaw made the symbols visible to subjects under certain conditions. However, this was quickly recognized by Rhine so such cards were never used in formal tests.[3] The charge of cherry-picking is not only untrue, it makes no sense. What C.B. describes would create a chance sequence of hits at the start and end of a run of Zener card calls, while in reality the hits tend to be *more* numerous than anticipated by chance, fall away, and then *recover* at the end of a run—a U-shaped curve. But in any case, had Rhine and his colleagues only accepted portions of an inspected run that by chance showed an excess of hits, the general effect would be randomly patchy, no sort of curve at all. Besides, this assumes that optional stopping is allowed, and that a tally of hits and misses is being kept during the process, neither of which was usually the case.

Since C.B. has been telepathic since the age of 13, he knows that he is lying—but many sf readers probably do not. So this unexpected reversal of what is often assumed to be the case—that psi does not exist, as he insists—provides a small shock when it is revealed. Even so, C.B. assures Briddey that other kinds of psi are simply non-existent—there is no precognition, no clairvoyance, no telekinesis, etc. The narrative supports these assertions. It is a way for Willis to keep a grip on the diegesis, the unfolding plot, the very contrary of Blish's accelerating and ever-more astonishing catalogue of psychic abilities revealed by Danny Caiden and his friends and foes in *Jack of Eagles*, the very model of a classic psience fiction story. Still, for her purpose—the exploration in zany madcap plotting of lies, evasions and socially masked anger and lust—this restraint and single theoretical focus works appropriately, if somewhat tediously.

Ψ

Even by itself, mind-reading provides Briddey with overwhelming threats. Refusing C.B.'s informed and hard-won advice at every turn, she attends an upmarket play with the increasingly loathsome Trent but is abruptly inundated by the tsunami of raw unfiltered thoughts from every side. An emcee calls for everyone to switch off their phones before the curtain rises; Briddey "hears":

> "Fucking rules!" someone said disgustedly, so loudly and so nearby.... *Yeah, and miss half the goddam show! ... It's a fucking dictatorship! ... I didn't pay two hundred dollars so some fag can tell me what I can and can't do....* [S]imultaneously, a female voice said, *he's really hot ...* and the blind-date woman said, *I don't care how old the theater is, I just want this date to be over!* But even though all three were speaking, they didn't mask one another like spoken voices would have [p. 192].

In Willis's version of psi, telepathy from a multitude is just like hearing linear speech but without mutual interference (not a feature of actual reported telepathy). Propinquity enhances vulnerability, driving Briddey to flee to the women's room to collapse, and C.B. to rush to the theater and her rescue. The sensory overload of mental uproar is unremitting, paralyzing. If psi were actually like this, those especially open to its assault would risk madness or suicide and need to find ways to insulate their minds from the unintended mental pollution from every side. Luckily, C.B. has long since developed "panic room"-style defensive barriers, and for many pages we follow Briddey learning the tricks of psychic self-defense.

Ψ

Arguably the driving impulse of this long novel is its pursuit of deceit and almost carnivalesque reversal of candor, which is found everywhere, often benignly or defensively intended. Briddey is constantly saying "No" while thinking *Yes*, or vice versa, and sometimes this spills over into what are probably small but telling japes on Willis's part: "Briddey" is a bride to be (or, as the events unfold, not to be) but also an analogue of Bridey Murphy, supposedly a 19th century Irish maid who was reincarnated as the Midwestern American Virginia Tighe; this was later argued to be a concoction of scarcely remembered childhood events, hypnotic "regression," and perhaps deliberate fraud.

Similarly, the usually incommunicative Jew C.B. Schwartz is really Conlan Brenagh Patrick Michael O'Hanlon, adopted by his Jewish stepfather. But in another sense, "C.B." recalls the Citizen's Band broadcasting technology prior to cellphones, which is an analogue of telepathy that everyone tuned in can hear, and "Schwartz" or "black" indexes to his status as "black Irish": "he had the classic dark hair and black-lashed gray eyes 'put in with a sooty finger.'" Trent Worth, meanwhile, is plainly worth*less*, another lexical reversal. With role reversal as a skeleton key, it is easy to anticipate the roles of put-upon young Maeve and the apparently pestish Aunt Oona, if not the wholesale voice-blocking and supportive role of the members of the Daughters of Ireland (p. 479) and their own paranormally gifted daughters and sons.

Inaccurate as Willis's treatment of psi functioning is, the book interestingly explores some of the possible consequences of telepathy if it ever gets fully switched on in a human brain. More than that, it reminds us, with a certain comic glee, how duplicitous we all are, in our own interest and for the perceived benefit of others—the "white lie," and economy with the truth, that serves as a social emollient. If there is any realism in this account, it might help explain (as the novel suggests in passing) the gradual extinction of psi faculties or rather the evolutionary emergence of the rest of us, those with haplotype genes for the inhibition of psi, most of the time.

43. Two Novels by Psychics (1978, 1998)

All the books and stories discussed so far have been created by professional or avocational writers. Is there any psience fiction written by actual psychics or parapsychologists, akin to the thriller novels written by forensic specialists (Kathy Reichs) or retired spies (John Le Carré)? Granted, numerous volumes have appeared about real world psi efforts, by the individuals involved in them, notably several by former military remote viewers Joseph McMoneagle (2002) and Paul Smith (2005), and by physicist Edwin May, scientific director of that program for many years, co-editor of the magisterial four volume *Star Gate Archives* (McFarland, 2018 and forthcoming).

The closest to psi *fiction* by practitioners, though, seems limited to two novels, to date. One is *Star Fire* (1978), a rather overwrought transcendentalist fantasy about psychic warfare and the emergence of a rock star psi god, by the late Ingo Swann. He was one of the prime instigators with Russell Targ and Dr. Hal Puthoff in originating what became Star Gate.[1] The second, from two decades later, is a surprisingly well-written novel by spoonbending Israeli psychic performer Uri Geller. *Ella* (1998) follows a British female version of his life, as the rhyming title suggests, but with plenty of skews introduced.

Ψ

Ingo Swann, a major developer of psychic remote viewing across time and space, urged us to consider

> the concept that perhaps it MIGHT BE DISADVANTAGEOUS to DEVELOP the superpowers of the human biomind.

You see, even ONE developed, truly developed, intuitive, telepath or mind-reader might shift the balances and parameters of all games played in the World—especially if those games are idiotic and senseless to begin with. (And this concept was the theme of my novel, *Star Fire*, published in 1978 by Dell.)[2]

A volume edited by Swann less than a decade prior to the collapse of the Soviet Union might cast a skeptical light on Ingo's precognitive prowess: *What Will Happen to You When the Soviets Take Over?*[3] Still, the political world of *Star Fire* retains an alarming similarity to some aspects of today's unruly planet with militaries and corporations threatening the use of ghastly new weapons to gain or maintain power, wealth and status at whatever cost.

Daniel Merriweather, a young singer-songwriter from the wrong side of the tracks, has the Dylanesque power to capture audiences with the beauty and challenge of his lyrics, bringing him great wealth. At 25 or so, this is not enough; he seeks higher truths, especially after he sang one day to a trillium flower and the flower sang back. His psi powers are unfolding, drawing him in directions he can't yet map clearly although the fields of his mind expand to cover the entire world, and then beyond: "he experienced floating in space above Earth, moving about the planet, perceiving scenes in distant places" (p. 83). "Earth itself seemed to recede, and he seemed to be unified with systems of stars far from the small planet circling its yellow sun, to become a consciousness in and *of* space itself ... the region in which he existed seethed and roared with energy (p. 65).

A lot of this recalls Ingo Swann's own reported capacities, such as a mental trip to the vicinity of Jupiter. He described details that Carl Sagan and other astronomers mocked as unrealistic but which were allegedly confirmed by the subsequent Voyager 1 space mission, such as a ring around the gas giant world. Careful scrutiny of his brief report suggests that the so-called ring was a scattering of crystals *inside* the atmosphere of Jupiter, nothing like what the spacecraft photographed: "Very high in the atmosphere there are crystals ... they glitter. Maybe the stripes are like bands of crystals, maybe like rings of Saturn, though not far out like that. Very close within the atmosphere (cited in Targ 2012, p. 32).

On the other hand, Swann's sketch (shown in Targ 2012, p. 33) does present one circle with two lines across the equator, obviously Jupiter, and another circle with a traditional Saturn ring extending on either side. Is this Jupiter with ring, not to be confused with the glittery crystals inside the atmosphere, or just a comparison of how Saturn would look from the same angle? Such are the hazards of interpreting remote viewer records.

Merriweather's experiences are far less ambiguous and liminal. To his horror, he stumbles on two vile secret mind control projects by the U.S. and the USSR, Tonopah and Tolkien, that might ruin the human future. Anonymously, he threatens to publicize their existence and purpose unless they are shut down. His psychic searches (indicated in scattered chapters by "Probing!") reveal that his earlier bout of tedious testing by a rigid organization that sounds like a cartoon blend of the Rhine establishment and Star Gate has left documentation that could lead government or criminal agents to him. So with a judicious application of psychokinesis from a distance, he causes all the paper records, and some stray flowers, to disintegrate into dust. Merriweather is starting to feel his oats, until he arranges in advance with his baffled agent to have his corpse cared for and then carefully throws himself under a taxi and is rather mysteriously preserved as Neomort 25-A.

Hundreds of pages of pre–Dan Brown shenanigans later, postmortem 25-A's exteriorized consciousness diverts two triads of military satellites, one set bearing the U.S. Tonopah horror weapon and the other the USSR Tolkien, and obliterates Omaha and Novosibirsk in noir set pieces, then wipes out an Azores island hosting a vile arms merchant. Finally he brings his own preserved corpse back to life as Sirius, a transcended entity, revealing this transformation to General Harrah Judd and Dr. Elizabeth Coogan, his tenacious pursuers.

> Sirius sang. His lips did not move.... When the song died away, Judd and Coogan started. The glow had faded. And with it the overpowering sense of an awesome presence. A slight young man stood before them, unusually handsome and serene, but otherwise perfectly ordinary, looking at them with a half smile [p. 361].

Sirius is psience fiction's *Homo superior*, a sort of super Ingo. His step by step transition to this estate throughout the novel makes *Star Fire* more than a Mary-Sue wish fulfillment for Swann,[4] a former "Operating Thetan VII" Scientologist, but perhaps only just.[5] The same might be said of Gully Foyle, extravagant unfolded superhero of *The Stars My Destination*—but then Alfred Bester never considered himself a potential Übermench. Perhaps in his Mark Twain corn pone mystical way John W. Campbell might have held hopes to prepare the way for such an evolutionary leap, but he never really supposed he was going to do the leaping. On the other hand, neither he nor Bester ever flew out of the body through interplanetary space to Jupiter and described its previously unseen rings.

Ψ

The *Publishers Weekly* (February 1, 1999) review of *Ella* offered Uri Geller an unexpected endorsement: "though his original motive may have been to advocate his beliefs, the result is a most respectable piece of storytelling." This is justified. Whether the reader regards Geller as an out-and-out charlatan[6] or a flashy but misunderstood psion, he proved to write well and compellingly about a Bristol teenager whose life was nothing at all like his own.

Poltergeist phenomena (a.k.a. "recurrent spontaneous psychokinesis") have traditionally been associated with troubled teens undergoing puberty but thwarted by social tensions, difficulties with parents and peers, or even bullying and persecution of the kind serious enough sometimes to lead to physical self-harm or even suicide. In commercial mass-market fiction, this is the territory of horror novels such as Stephen King's *Carrie* and Colin Wilson's psience fictional *The Philosopher's Stone* with its exuberant PK effects (see Chapter 32, above). Such paranormal outbursts apparently do occur, and have been studied extensively by parapsychologists such as the late Dr. William G. Roll and one case in particular investigated by Guy Lyon Playfair.[7]

Ella Wallis, on her 14th birthday a classic instance of this syndrome, is the psychologically wounded child of brutish, philandering, religious hypocrite father Ken and bitter, disappointed French mother, Juliette, who secretes bottles of gin all around the house. Shy, distracted Ella is regarded as a dimwit, and is. For her birthday a large and rather sickening cake has been provided by her Auntie Sylvie (a former child poltergeist focus, it turns out), which at an especially fraught moment rises from the table and flings itself on the carpet. Of course Ella is blamed; she denies it until her father seizes her cruelly. She wakes in the night floating above her bed, and her drawing of an angel, posted on the wall, is a charred remnant (an event that is never explained). So much for the opening chapter. Unlike Ingo Swann, Geller gets right into the action.

The following day, her fury at being mocked by a repressed and bitter teacher results in the explosive destruction of a heavy glass window in the class door. We are clearly and rather quickly in major poltergeist territory. Within days, Ella is condemned as a thief when a stolen model car is found in her bag (it's a fair cop: she presumably teleported it unconsciously). Her class and its teacher are in uproar as objects hurtle about, lights flare, Ella's desk flings itself into the air and smashes amid whining shrieks, and a Christmas nativity manger bursts into flame except for the white-swaddled Infant Jesus. Naturally Ella is accused of throwing lighted matches and sent packing. It is a nice presentation of how the urge to normalize the preposterous might allow sensible people to deny what they all have just witnessed (as in Asimov's "Belief").

Ella's unpleasant and probably child molesting uncle Robert, lay preacher of the Pentecostal Church of Christ Reborn, decides to exorcize the demons from her. As he babbles pieties and fondles her breasts, Ella slams one sharp knee into his engorged groin and floats up between holy candles before rushing to her bedroom for her first bloody period. Incautiously, Robert recounts this exorcism in the church pamphlet *Shout for Joy*, and mid–40s journalist Monte Bell seizes the morsel and writes up Ella's case for the local paper (which appeals to his younger and snooty editor even though the story fails to report Ella "bending spoons or something" (p. 96). Geller is wry.

Finally, handsome paranormalist journalist Peter Guntarson, a DPhil from Oxford in psychology and parapsychology, understands that these astonishing occasions are due to psi rather than demons or angels, although the ambiguity remains beyond the last page. Amid the domestic and media monsters, he alone has compassion for this sad, challenged child, as well as a burning eagerness to comprehend her gifts. He triggers apports (a cup appears in the air, smashes into three pieces), telepathic contact between the two of them, and levitation that is photographed and used to launch Ella's fame.

Soon she is sought after by a TV host who despairs of her numb silence—"I mean, who's she supposed to be, fucking Helen Keller?"—until she slumps in her chair and flies it thirty feet into the heights of the studio. Swiftly a Portuguese public relations superstar, "Dr." José Miguel Dóla, combines with her money-hungry oaf of a father to entrap Emma into a comprehensive contract locking Dr. Guntarson out. For mutual advantage, Guntarson and Dóla are soon secretly in cahoots, although not for long. All of this background rings disturbingly true, as does the increasingly intrusive media hysteria. Geller and his ghost writer (if he had one) know their way around.[8]

Amid the crises of her dysfunctional family, an abundance of colorful and rather alarming poltergeist events wax and wane. Peter announces his theory that psi is the expression of an Intelligence, and he is the man to deal with it, having an IQ of 183 as well as a close protective bond with Ella:

"There is a guiding intelligence at work here When you are shown a sign, why dismiss it as coincidence? ... I have been assigned. As Ella's interpreter.... Psychic power has its own personality. It can pick and choose people, order events, just like we can. Being psychic isn't just the obvious stuff—levitating or whatever. It is being in possession of a non-human intelligence.... Some people choose [the alien] explanation, and some people choose the God explanation.... I am her enabler. Her catalytic converter, the key that unlocks her.... Whatever metaphor you choose, you have to understand that without me Ella is just a random psychic generator. She can't control it. She can't understand it" [pp. 202–03].

Pretentious and grandiose as this sounds, it seems from what we see to be the truth—except that when her beloved brother develops an inoperable brain tumor poor low IQ Ella chooses for herself. She demands access to a TV interview, and by now her levitations have made her hot news. Ella calls for prayer in aid of Frank, and all across Britain weeping viewers concentrate their psi intentions to heal not only her brother but hundreds or perhaps thousands of others. It is a version of Geller's own TV stunts where he bent a key or restarted a dud watch and viewers found the same marvel replicated in their kitchens.

For Ella, this has been a risky move. Still besotted by the Nordic god in riding leathers, she awaits her 16th birthday so they might be legally married. Guntarson is hardly enchanted by this prospect, but uses her devotion to establish a worldwide Ella Center of prayer, and the money rolls in. Guntarson is vilified in the press, and finally chooses a *coup de theatre* in which he drowns himself as a huge display of faith, certain that Ella can resurrect him. This doesn't happen; having watched his spirit ascend, she chooses to let him remain in heaven. At last waif-thin Ella, who has not eaten for months, flings herself from the top of Mount Sinai, but we surmise that this is not in despair. Her most devoted fan recalls "that at the instant Ella Wallis walked off the path, she grew wings and flew into the void as an angel" (p. 435).

Rather as Ingo Swann's *Übermench* Sirius glowed, Ella is effulgent in moments of rapture and prayer:

> Ella's head and shoulders were radiating light.... The effect grew by the second. Ella was surrounded by brilliance. Light shone from her silver hair like strong sunlight off the sea.... Her face was upturned, and with a sigh she lifted from the cushions into the air.... All the light came from Ella. The shimmering glow surrounded her like a star blazing beneath her skin. Like a fiery orb descended from the heavens upon her. Like a halo [p. 296].

Geller is drawing here upon sacred art and iconography, and legends of saints in all their humble pomp. He has never made such claims for his own powers, which seem restricted to metalware, mentalism and mysterious information he obtains for rich corporations and perhaps the Israeli armed forces. He is not known to levitate.

44.
Short Stories (1960s–1990s)

Poul Anderson, *"Night Piece" (1961)*

Although telepathy between humans and aliens had been established as a psience fictional trope at least as far back as "Doc" Smith's Lensmen series, it seemed to have a revival near the beginning of the 1960s, perhaps as the ultra-conservative postwar mood slowly gave way to the delight in novelty and transgression of what would become known iconically as "the Sixties." Clifford Simak's *Time is the Simplest Thing* (1961) conveyed with distaste both the general authoritarian revulsion for mental miscegenation and cultural exogamy, with an alien Pinkness infusing itself into the very being of psi-powered explorer Shepherd Blaine. We shall see varieties of this device unfold in stories from the 1970s onward.

Some of these fictive explorations assume a rough parity between the communicants at either end of the mind-call. Others take the more frightening and justifiable view that minds beyond the human will almost certainly be enormously more powerful, and with their own interests, than ours. The editor of the *Magazine of Fantasy & Science Fiction* in 1961, Robert F. Mills, made no bones about what Poul Anderson intended in this story. His opening blurb states: "As a mouse knows about cheese and traps but does not really know about Man, is it possible that Man knows about, for example, magnetic fields, but knows nothing about the Superior beings of whose existence those fields might be an evidence?" (p. 56).

A nameless and Faustian researcher has built an experimental ESP amplifier, which he fears has "opened hell gate" even though he refuses, as a rationalist, to believe this (p. 57). What he does believe, or at least suspect in his terror, is that Superior exists on Earth—not *Homo superior* but some adjacent fork of evolution, maybe split from ours at some simian stage or farther back, much farther, "perhaps as far back as some amphibian in the Carboniferous" (p. 64). What would Superior be like, compared to humankind? More muscular, with keener vision, better reasoning powers? No.

> Shouldn't he rather posses only a modicum of reasoning ability by our standards, very weak instincts, a few reflexes, and no tropisms? But his specialty, his characteristic mode, would be something we can't imagine…. Conceivably in the ESP field—Now I'm letting my hobby horse run away with me again. (Damn it, though, I *am* starting to get reproducible results!) [p. 63].

In a rush of somewhat overwritten poetry of extremis, Anderson takes his hapless researcher through a sort of exaggerated biological epic of evolution, experiencing life in the ocean, below the thrashing spasms of lightning at the roiling surface, shuddering at the muddy floor, watching aghast as the subterranean crust rises and surges through the clouds. It is a death struggle between different factions or instances of Superior. Huddled in a bar over a drink, he knows that these visions are "merely the way the forces, the current, felt to him. Frantically seeking a balance, his mind interpreted these unnatural stimuli in the nearest available human terms" (p. 67).

But if his technical tampering with ESP amplification has brought him to this lethal awareness of Superior, why are routine parapsychologists spared this Grand Guignol? Maybe his fate "was a most improbable accident. Other men could still go ahead and study ESP phenomena as much as they cared to, learn a lot, use their knowledge, all in perfect safety, with never a hint that on a higher level of those phenomena Superior carried on huge purposes" (p. 64).

What Anderson and his nameless, brave scientist are toying with here in this little horror story (and it is quite horripilating, causing the skin to crawl) is the archaic dread of demons and alternative child-gnawing gods. Self-professed Magickal sorcerers such as Aleister Crowley, known as the Great Beast, and other would-be tamers of a supposed Satanic realm are familiar with this imagery, as James Blish displayed quite remarkably in his two volume sf-cum-fantasy novel *Black Easter/The Day After Judgment*. Amazingly, it was precisely such superstitious fears that led a number of highly placed U.S. military and Congressional Christian fundamentalists to force the closure of the Star Gate psi program.[1] Unless, of course, it was the directing impulse from Superior...

Robert Silverberg, *"Something Wild Is Loose"* (1971)

In early science fiction short stories, telepathy was very frequently used as a device to allow bold human explorers to arrive at the hospitable planet of a distant star and immediately communicate with (generally humanoid) aliens. Sometimes translating machines were invoked, but they seemed both distancing and rather implausible. (Oddly, recent "deep learning" computer systems might already have gone part of the way to resolving this problem with interstellar aliens, so long as the to-and-fro learning can be conducted en route, to save a lot of time sitting around incommunicado waiting for the computers to learn each other's codes.) This rather cinematic Silverberg story, somewhat resembling James White's much loved Sector General space hospital series, has just one low-grade telepath available to deal with a plague of fearsome and deadly nightmares triggered by a small invisible hijacked alien who just wants to get home.

Dr. Peter Mookherji is a neuropathology resident in Long Island Starport Hospital when the Vsiir arrives from the cargo hold of a starship disrupted during its return to Earth by the screaming horrors afflicting its crew. The intelligent alien has been attempting to enter minds apparently blocked against it, touching them with the tip of its mind: *Please listen, I mean no harm. Am nonhuman organism arrived on your planet through unhappy circumstances, wishing only quick going back to own world.* Almost immediately its attempted contact shocks two critically ill patients to death. Despite being a telepath himself, Mookherji ridicules the suggestion that something wild and mentally intrusive is loose, preferring potential diagnosis of virus, encephalitis, chronic hallucinations.

Meanwhile, he struggles without success to entice from deep coma a young woman whose family died during an accident on Titan. He can enter the unconscious mind of teenaged Satina and tempt her with glowing fantasies of sun and surf, free flight with "a floater-node fastened to her belt, lifting her serenely" (p. 29), but her resistance has rendered his efforts futile.

The core of the psience fictional aspects of the tale are found in Mookherji's heritage and tentative grasp of the dynamics of psi:

> The gift had been in the Mookherji family for at least a dozen generations, helped along by shrewdly planned marriages designed to conserve the precious gene; he was more adept than any of his ancestors, yet it might take another century or two of Mookherjis to produce a really potent telepath.... [M]any members of his family in earlier times had been forced to hide their gift from those about them, back in India, lest they be classed with vampires and werewolves and cast out of society [p. 28].

How does psi work in this fictional universe? More like memorizing an entire difficult play than have a cheerful chat, it appears. "There was nothing easy or automatic about his telepathy; 'reading' minds was strenuous work for him.... As far as anybody knows, telepathic impulses propagate somewhere outside the electromagnetic spectrum.... God help me if I'm wrong, Mookherji thought, far below the level of telepathic transmission" (pp. 28, 39, 42).

The Vsiir, exhausted, alone and increasingly hopeless, hears a call: "Clear, intelligible, unmistakable. *Come to me.* An open mind ... speaking neither the human language nor the Vsiir language, but using the wordless, universally comprehensible communion that occurs when mind speaks directly to mind" (p. 42).

Mookherji gambles, takes the choice that wrecked the budding romance in Poul Anderson's "Journeys End":

> He opens his mind wide.
> The Vsiir rushes into Mookherji's brain [p. 45].

C.J. Cherryh, *"Cassandra"* (1978)

Whether myth, legend or some residue of real prophetic history, Cassandra's tale is one of the ancient warnings against the exercise or at least the announcement of precognition. It's not a matter of "Don't wish for what you want, because you might get it," but rather ominously "Don't look into the future, because it is awful and everyone dies at the end and nobody will believe you but they will certainly hate and shun you." This little eight-page piece by Caroline Cherryh, which won her the 1979 Hugo for best short story, is a sort of prescience companion to Walter M. Miller's "Command Performance," where telepathy led first to danger and attack and then to appalled solitude.

This narrative is bleak and even briefer than Miller's: Alis, known to everyone in town as "Crazy Alis," lives in an experiential world of layers, ghostly presences of fire and insubstantial ruin and terminal disruption and death but without flaming heat or the stench of corpses. Does she know this is the post-apocalypse future she sees? Do others ever come to recognize the fateful validity of her reports, not that she offers any now, having learned her sibylline lesson? A young man from out of town speaks to her when she dashes after him: "scuffed brown leather coat, brown hair a little over his collar.... A young face, flesh and bone among the ghosts" (p. 201).

The spectral flames cease. He buys her dinner at pricey Graben's, tells her she is beautiful. One images a sort of *Max Max*-meets-Pre-Raphaelite loveliness, with haunted eyes. Sirens shriek, finally people rush like Gadarene swine to their doom. Alis drags Jim to safety in a café cellar. She cannot tell him. Ruined bricks fall from an unstable wall. Jim is crushed into ghostly death. A looter menaces her, at last, and she waits until his foretold doom has passed and retrieves her stolen cache.

> One could live in ruins, only so the fires were gone.
> And the ghosts were all in the past, invisible [p. 206].

This small parable must have stung many readers forty years ago, the Soviet war machine and the American war machine still menacing each other with nuclear doom. Today the dread is less distinct, and perhaps too large and barely visible: others with nuclear weapons and missiles to throw them, true, but also the hothouse effect slowly preparing to burn the life of the world in a hundred or a thousand years, or maybe fifty years, or sooner. Can precognition aid us in this moment of endless dread? Or can psience fiction, and perhaps psi fact, provide no more exact message than that of poor demented Alis in the cold ruins?

Brian M. Stableford, *"The Oedipus Effect" (1991)*

Simon Sweetland is a theorist of paranormal Talents, although lacking any himself, Head of the Department of Paranormal Resources Coventry with a single assistant. Sweetland is one of the few dozen in the British scientific civil service studying the topic. Perhaps eventually they would find an explanation for Talents. He meets with Lewis Fay, Senior Claims Adjuster, Midlands and West, of Family Provident. Fay has read his recent paper, "A New Interpretation of the Precognitive Paradox," in the *Journal of Paranormal Studies*, and wishes to hire Simon as a consultant.

A small percentage of the Paranorms in the community are able to deliver valid prophecies, and that immediately produces the obvious paradox.

The likelihood of the event happening is altered, which Simon has dubbed "the Oedipus Effect." The classic prediction that young Oedipus would one day slay his father and marry his mother is what triggers the improbable chain of cause and effect bringing about exactly this result. Simon admits ruefully that it is not a very apt coinage, because most precognitive future-telling turns out to negate the prophecy rather than bring it about. You foresee yourself perishing in a gruesome plane crash, so you cancel your reservation and the prophecy is refuted. That doesn't mean the plane failed to crash—unless you warn the airline and prevent the flight—just that you are not one of the victims.

How is this accounted for? As Philip K. Dick realized years earlier, pre-cogs do not see a determined future but one from an array of possible outcomes. Simon, however, has argued that this is a fake solution. Really it is simpler to suppose that what seems to be precognition is actually an unconscious warning of a psychokinetic act bringing about just that outcome. (This has long been a well-known line of thought in parapsychology, but Sweetland and his fellow experts have apparently not been keeping up.)

Family Provident's interest is aroused by the death of one of their insured, Thomas Hemdean, predicted by his eight-year-old son, Nicholas, in front of witnesses including Alex, a neighbor, and his daughter. Nicholas had been tested by DPR but dismissed as lacking Talent; if this deterred his father from acting on the prediction, the company might be liable for considerable compensatory losses. But if it can be shown that the boy might unconsciously have *caused* the death of his father by PK—an ironically genuine Oedipal effect—this would spare Family Provident a large payout and set a useful precedent.

A tabloid newspaper shows a photo of the boy: "blond, thin-faced and unsmiling, somehow rather furtive, he looked completely unlike his dark-haired, chubby and seemingly self-satisfied father." An old joke comes instantly to the suspicious reader's mind, although not to Simon's. It goes like this: *A palm reader at the carnival tells a boy his father will die tomorrow in an accident on his way to work. Wailing, the child tells his father this prophecy, and for the sake of peace the man agrees to stay home, under scrutiny from the entire family. By nightfall of the following day, the father can't stand being locked-in any longer, and opens the front door. A dead man lies sprawled across the front steps. He is the milkman.*

Fifteen pages after they and we see the photo, the neighbor, Alex, is similarly killed in an accident with some similarity to Nicky's dream. In the words of another old joke: Oedipus, schmoedipus.

Conclusion

We began this discussion by noting three major inputs to science fiction in the mid 20th century, prior to the astonishing acceleration of computers and their apps toward the end of the millennium: space flight, the atom, and paranormal powers. To date, a number of informed books have reported on the literary evocation of both space travel and nuclear holocaust (or, more rarely, nuclear benefits, however ambiguous). As far as I am aware, there has been no equivalent volume, until now, devoted to the varied science fiction based on parapsychological phenomena.

The primary aim of this book has been to fill that fundamental gap—to offer an insight into the numerous forms of imaginative engagement with the paranormal that constitutes what I call "psience fiction" rather than, say, horror fiction with paranormal aspects (Stephen King's *Carrie* or *The Shining*, for example) or pure fantasy. Many monographs have looked closely at the work of individual sf authors, sometimes briefly teasing out the parascience either implicit or explicit in some of the texts. Solid critical work has flensed the novels and stories of James Blish, Samuel R. Delany (some of the best theorized critique written by Delany himself), Philip K. Dick, Frank Herbert, Ursula Le Guin, Joanna Russ, Theodore Sturgeon, Roger Zelazny—but each of these studies takes a global approach to the diverse frames and tones of these masters. The model I have pursued here is closer to that pioneered by Gary Westfahl in *Islands in the Sky: The Space Station Theme in Science Fiction* and *The Space Suit Film: A History*.

This approach, essentially allowing currents of thought, topos and trope to display themselves in historical and frequently mutually interactive dynamics, quite deliberately declines to engage closely with any specific critical theory or, on the whole, even semiotic disciplines such as psychoanalysis, except when they are built into the plot (as, in particular, Alfred Bester's *The Demolished Man*).

Entire university library shelves might be set aside for future studies investigating or interrogating these psience fictional texts under the searching gaze of specialists in Lacanian or Marxist or libertarian analysis, or the ontological skepticism of Baudrillardians, or indeed ethologists, genomics specialists, cognitive neuroscientists, hedge fund theorists, futurists and semanticians. At least some of these future forays will benefit by paying attention to the books and short stories selected here for their predominant characteristics as psience fiction.

As is obvious, most of the narratives examined here have been presented in their order of publication, which to some extent reveals clumping that reflects popular taste or editorial quirk. It's necessary to stress that almost all popular fiction has traditionally taken months or even several years to appear in print (unlike the kind found today online, where a story can appear moments after the writer has keyed in the final letter).

What's more, while some writers such as Robert Silverberg, Barry Malzberg and Isaac Asimov were fantastically fluent and wrote fast, producing publishable copy in a matter of days or a few weeks or a couple of months at most, plenty of writers require years of painfully determined slogging on the path from first notion or thematic tickle in some buried portion of the brain through to the final unfolding, reshaping, honing, polishing, submission and acceptance by a publisher. So a simple date attending each title need not imply that each was open to influences current at that time, or even a year earlier. On the other hand, sf writers are notoriously communal in their practice, sharing ideas over drinks or under the sheets, attending master classes where they show each other their works in progress, often unafraid to criticize each other's precious product in a way they would never dare slight spouses or children.

Bearing all these cautions in mind, it is still apparent that the influence of John W. Campbell's crusade in favor of psi, both in editorials, articles, serials and stories he often all but sketched out in advance, yielded a large number of fictions with this general family resemblance in the early to mid 1950s. Except for constraints of space, quite a few additional titles could have been added to the Table of Contents. Still, it is evident that just in my selection (and in some cases leaving aside sequels) psience fiction offered six titles between 1935 and 1949 (15 years), 17 between 1952 and 1959 (eight years), 12 between 1962 and 1969 (eight years), eight between 1970 and 1976 (seven years), three between 1978 and 1987 (nine years), one between 1988 and 2010 (23 years) and three between 2011 and 2016 (six years).

So more than twice as many of these psience fiction novels were serialized or first published as books in the 1950s and the 1960s as in all the other years combined, and then the proportion of at least moderately significant books of this kind sagged, with an astonishing Quiet Zone of 23 years prior to 2011. Some of this was surely author exhaustion or reader boredom after the Campbellian deluge. Some resulted from competition by fresh topics of

Conclusion

scientific or narrative interest (especially the rise of cyberpunk from the early 1980s and then the slow awareness of a possible technological Singularity on the medium-near horizon spelling doom to any traditional set of sf tropes). Meanwhile, fantasy overwhelmed sf in the marketplace, often absorbing versions of psi tropes in the form of magic or sorcery or steam-punkery, which snatched the market footing out from under serious fictional work drawing on the real psi research that ironically and belatedly was being conducted in a few universities and in secret labs funded by several nations.

Let us consider the principal components of these books and stories by chapter order and roughly in order of publication, where TP is telepathy, PK is psychokinesis, PRECOG is of course precognition or prophecy, and VARIOUS or + indicates promiscuous blendings of psi varieties:

1. Donald Macpherson (George Humphrey), *Go Home, Unicorn* (1935): PK
2. Olaf Stapledon, *Odd John* (1935): TP +
3. E.E. Smith, *The History of Civilization* (1937/1951): TP +
4. A.E. van Vogt, *Slan* (1940/1946/1968): TP
5. James Blish, *Jack of Eagles* (1949/1952): TP +
6. James H. Schmitz, *The Witches of Karres* (1949/1966): TP + TELEPORT
7. Alfred Bester, *The Demolished Man* (1952): TP
8. Zenna Henderson, *The People* (1952): TP
9. J.T. McIntosh, *The ESP Worlds* (1952): TP
10. Theodore Sturgeon, *More than Human* (1953): VARIOUS
11. Mark Clifton and Frank Riley, *They'd Rather Be Right* (1953–1956): TP
12. Mark Clifton, "What Thin Partitions" to "Remembrance and Reflection" (1953–1958) TP +
13. Wilson Tucker, *Wild Talent* (1954): TP
14. James H. Schmitz, *The Ties of Earth* (1955): TP +
15. John Wyndham, *The Chrysalids/Re-Birth* (1955): TP
16. Robert A. Heinlein, *Time for the Stars* (1956): TP
17. Frank M. Robinson, *The Power* (1956) *Waiting* (1999): TP +
18. George O. Smith, *Highways in Hiding* (1956): TP + PERCEP
19. Alfred Bester, *The Stars My Destination* (1956–1957): TELEPORT + TP
20. Lan Wright, *A Man Called Destiny* (1958): ANTI-PSI
21. Marion Zimmer Bradley, *Darkover* series (1958): VARIOUS
22. Jack Vance, "Parapsyche," "The Miracle Workers" and "Telek" (1958): PK
23. Short Stories (1940s–1950s):

Conclusion 187

Katherine MacLean, "Defense Mechanism" (1949): ANTI-PSI
C.M. Kornbluth, "The Mindworm" (1950): TP
Walter M. Miller, Jr., "Command Performance" (1952): TP
Isaac Asimov, "Belief" (1953): LEVITATION
Algis Budrys, "Riya's Foundling" (1953): TP
Cordwainer Smith, "The Game of Rat and Dragon" (1953): TP
Brian W. Aldiss, "Psyclops" (1956): TP
J.T. McIntosh, "Empath" (1956): EMP
Poul Anderson, "Journeys End" (1957): TP

24. Mark Phillips (Randall Garrett and Laurence M. Janifer), *Brain Twister*, *Impossibles* and *Supermind* (1959–1961): VARIOUS
25. Arthur Sellings, *Telepath* (1962): TP
26. Keith Woodcott, aka John Brunner, *Crack of Doom/The Psionic Menace* (1962–1963): TP
27. John Brunner, *Telepathist/The Whole Man* (1964): TP
28. Dan Morgan, *The Sixth Perception* series: (1967–1975): TP
29. Richard Cowper, *Breakthrough* (1967): TP +
30. Anne McCaffrey, *Talents Universe* (1968): VARIOUS
31. Philip K. Dick, *Ubik* (1969): PRECOG +
32. Colin Wilson, *The Philosopher's Stone* (1969): VARIOUS
33. Joanna Russ, *And Chaos Died* (1970): TP
34. Lester del Rey, *Pstalemate* (1971): PRECOG
35. Robert Silverberg, *Dying Inside* (1972): TP
36. Katherine MacLean, *Missing Man* (1975): TP
37. Robert Silverberg, *The Stochastic Man* (1975): PRECOG
38. Octavia Butler, *Mind of My Mind* (1976): TP
39. Joan D. Vinge, *Psion* (1982): VARIOUS
40. Lucius Shepard, *Life During Wartime* (1987): TP
41. Carrie Vaughn, *After the Golden Age* (2011) and *Dreams of the Golden Age* (2013): VARIOUS
42. Connie Willis, *Crosstalk* (2016): TP
43. Two Novels by Psychics (1978, 1998): LEVITATION +
44. Short Stories (1960s-1990s):
    Poul Anderson, "Night Piece" (1961): TP
    Robert Silverberg, "Something Wild Is Loose" (1971): TP
    C.J. Cherryh, "Cassandra" (1978): PRECOG
    Brian M. Stableford, "The Oedipus Effect" (1991): PRECOG

Clearly, the classic psi mode—"mind reading," "thought transference" or ESP/telepathy—is massively overrepresented in these selections, sometimes accompanied by poltergeist/telekinesis/psychokinesis effects, but rarely by teleportation of humans from place to distant place, let alone distant times.

An interesting version of teleportation was Steven Gould's *Jumper* (1992), *Reflex* (2004), and *Impulse* (2012); these, somehow, while exciting and amusing, do not seem part of the core psience fiction canon in the exultant way Bester's *The Stars My Destination* does.

More psi novels played with alternatives to the classics, and some became classics in their own right: Clifford Simak's *Time Is the Simplest Thing* (*Analog*, 1961) opens with a frightened mind-teleporter who returns 5000 light years with an embedded alien blob, the Pinkness, who famously opened their telepathic conversation with *Pal, I trade with you my mind*. But the psi component here is really only a narrative device, for all its importance. Lloyd Biggle's *The Angry ESPers* (1961) is undoubtedly psience fiction, but of a routine kind, as is Alan E. Nourse's collection *Psi High and Others* (1967). Keith Laumer's *The Infinite Cage* (1972) and Daniel F. Galouye's *The Infinite Mind* (1973) make use of superpowers, perhaps to a fault. Worth a look are Herbert's *The Godmakers* (1972) and Gordon R. Dickson's *Analog* serial in the same year about a gigantic psionics machine, *The Pritcher Mass*, as examples of well-crafted but not especially influential tales.

Precognition, one of the favorite psi capabilities often used by Philip K. Dick, seems to enter more often into play as we approach the latter parts of the 20th century. It was also at the heart of Frank Herbert's famously best-selling *Dune* sequence starting in the 1960s, together with genetically engineered and spice-triggered multi-generational mnemonic recall—perhaps a kind of concatenated reincarnation—by the Jesuit-like Bene Gesserit Mothers.

Another serious attempt to map the future course of human history was Gordon R. Dickson's unfinished Childe Cycle, which parsed human cultural and psychological dominants into three major components (faith holder, warrior and scientist) and projected entire worlds commanded by one of these dominants or another. Beginning with *Dorsai!* (1959), under Campbell's influence, the sequence was threaded by psi powers, including levitation, combined with immense information storage and cognitive enhancement in its Final Encyclopedia, an archive mostly existing in some higher dimension. Exotic psychic and transcendent elements are crucial, but blended so thoroughly with others (as in the *Dune* series) that this long work is not best treated as psience fiction.

Some novels, again mostly not discussed above, also combine ESP with computer adjuncts, an aspect of advanced neurotechnology increasingly heralded by popular science articles with titles like "The Coming of Computer Telepathy!" and "Machines Can Read Your Brain!" An example was Roger Zelazny's and Fred Saberhagen's *Coils* (1982), which seems today more like cyberpunk several weeks ahead of William Gibson. These, too, are only marginally psience fiction—but perhaps one could say the same about Connie

Willis's *Crosstalk*, and certainly should in regard to her intelligent and funny novel about Near Death Experiences (*Passage*, 2001).

The worthy novels cited in the previous paragraphs do not form part of the discussion here because by some critical measure they seem remote from the main stem of psience fiction. That observation applies even more strongly to the first two books in our list, *Go Home, Unicorn* and *Odd John*, but those were early harbingers of a form only just then creeping from the seas and venturing on to the gritty sands.

Ψ

It is possible that in the first two decades of the 21st century, and beyond, we have been watching as the land creatures, after a brief century of exploration, carry back new recognitions to the ocean of story. Or perhaps the discoveries and practical applications of psi, being made by a small cadre of academic and military researchers, is almost ready to burst asunder the reigning paradigms of physics and biology. If that happens, we might find ourselves, to our astonishment and perhaps chagrin, living in one or more of the psience fictional worlds dreamed in that century of exploration by our ingenious and prescient writers.

This proposition, taken either literally or as a metaphor of discontinuous change, appears to be a central theme of psience fiction, often masked by the propulsive dynamic that reflects science fiction's beginnings as an adventure-story genre. It is a fiction intent on what we might call the biological trajectory toward enhanced consciousness. On a parallel course, many real world parapsychologists, like many mystics, are persuaded that capital-C Consciousness is the ground state of reality, from which emerge space, time, energy, information, and eventually individual minds all somehow connected by a mysterious and primordial entanglement. Philosophers call this hypothesis "panpsychism," in which everything in the universe has some small or larger ingredient of awareness and even consciousness.

(Personally, I find this postulate ludicrous and even unintelligible,[1] as grotesque as the long-maintained pre-microscope scientific belief that spermatozoa contain teeny little shrunken humans—rather than coded recipes—ready to be jump-started in the food store of the ovum.[2] But I can still enjoy science fiction where panpsychism seems to be the secret constituent, in the same way I enjoy time travel fiction despite its extreme implausibility.)

Ψ

As noted earlier, Alexei and Cory Panshin attempted to explain the source of science fiction's impulsive power in its frequent appeal to the marvelous, often in the form of a call to mystical "evolutionary" advancement. Indeed, the subtitle of their big history of the genre/mode makes that explicit:

The World Beyond the Hill: Science Fiction and the Quest for Transcendence (1989). Beyond any coarse power fantasies of psi as a tool for spying on the thoughts of others, flying without wings, foreseeing future stock market outcomes, healing or harming others by sheer mental determination, and so on, such a quest for transcendence is presented as salvific, the emergence of an improved and clarified humanity and the defeat of loneliness. It is no accident that this state is not uncommonly portrayed by an illuminated body or light-radiating eyes, often enough a golden glow emanating from the psi-enhanced (as in the climax of Ingo Swann's *Star Fire* and in key moments of Uri Geller's *Ella*). The auric iconography is familiar from many sacred or "wisdom" traditions; it might be argued that this, in turn, is derived from experiences with the psi-illuminated. Or not—such reports share a lot in common with rapturous moments brought on by brain changes triggered by ingesting LSD, psilocybin, ayahuasca, or "magic mushrooms,"[3] or undergoing a near death shock.

For the Panshins, the dream-like or oneiric character of early A. E. van Vogt sf is the very model of Modern (rather than postmodern) major science fiction. His work engages the reader at a level more akin to the psychedelic than anything familiar from westerns or detective fiction or romances, let alone most bourgeois "domestic" literary novels of inner and outer social life. Eventually, after his season with Dianetics, his further work degenerated into a jumble of random poorly written scenes, but at his best van Vogt did somehow convey a vision of lofty and magnificent accomplishment in a blazing future more like van Gogh's vortical multi-hued starscape than the latest engineering report from the International Space Station or vast solar energy array (wonderful though they are, in the way of Campbell in his technophilic *Analog* avatar).

For the next step, the Panshins identify the "Lewis Padgett" stories by husband and wife collaborators Henry Kuttner and Catherine Moore. This combinatory sf writer with two left and two right cerebral hemispheres delved unusually deeply into the unconscious for the Lovecraftian "sense of imminent incomprehensible strangeness trying to break though into conventional reality.... Transcendent mystery was capable of appearing anywhere at all, and it might involve the most ordinary of contemporary people" (pp. 591–92). How do we know it when we see it?

> Padgett's answer was that anything transcendent to us, wherever it is encountered, must have an appearance of bizarre strangeness, of irrationality, and that the way for humanity to bring itself into alignment with this higher aspect of existence must lie in the cultivation of non-rational thought processes [p. 592].

The Kuttners expressed this breach of the domestic and customary due to the hyper-van Vogtian in such mid–Second World War stories as "Mimsy

Were the Borogoves" (*Astounding*, 1943), where two small children find a hoard of teaching devices from the future and retrain their still-flexible minds. A psychologist explains to their parents:

> Let's suppose there are two kinds of geometry; we'll limit it, for the sake of the example. Our kind, Euclidean, and another, we'll call it x ... based on different theorems. Two and two needn't equal four in it; they could equal y^2, or they might not even *equal*. A baby's mind is not yet conditioned, except by certain questionable factors of heredity and environment [cited, p. 597].

The rewired children use the opening of Lewis Carroll's nonsense poem "Jabberwocky" as a map into these higher spacetimes, and vanish in a direction their father cannot comprehend. Years later, Mark Clifton's "Star, Bright" would adapt and somewhat re-domesticate this proposition, again for Campbell. At any rate, this story and some others akin to it would seem, for the Panshins, "to mark some sort of transition in the perception of transcendent non-rational consciousness.... Human beings might get along a lot better with the new transcendence if they would only learn non-logical modes of mentation more in tune with the actual nature of reality" (p. 600).

A decade or so later, John Campbell had assimilated this doctrine to the extent that his enthusiasm for psionics had replaced his brief flirtation with Hubbard's Dianetics, another kind of irrational would-be transcendence. But the shift the Panshins perceive might allow "utter strangeness [to] appear suddenly in our midst, and we also might find elements of familiarity at the most remote removes of existence.

"And the subject matter of this new myth," they assert, "would be transcendent consciousness" (p. 648).

Ψ

But is it entirely a new myth? Peter Lowentrout, who is well informed about science fiction (he is a past president of the Science Fiction Research Association as well as a professor of religious studies), offers an analysis of the metaphysics of psience fiction that combines elements from Heidegger and Jung. While he does not press the case for psi as proof of divine design, he argues that certain human experiences have been shaped by natural selection over hundreds of millennia into what amounts to inborn metaphysical archetypes. The way we tend to archive our interactions with the world, "chunk" them for swift recall and action, lends itself to gratifications that tend to be blocked or muffled in a desacralized post-communitarian world. Myths that evoke the old certitudes and hopes are those that persist longest and provide the deepest satisfactions—even in narratives like much of science fiction that seem even more mechanized and alienated than our own. Thus, he argues, the assumptions of the metaphysical SF psi story, that mind is real and grounded in a transcendent unity of being, speak to deep needs in us all, and for most readers a fictional world seems more complete with them than without, whether or not those readers are explicitly aware of their presence. The human species evolved in the context of such assumptions concerning itself and the world, and whatever one believes of their ultimate truth or falsity, thousands of generations of humans and proto-humans bear witness to their psychological reality. These assumptions ring true when used with skill in the fictional portrayal of people and societies. Too, no system is a self-contained whole, and a fictional world of epic proportions without a point of transcendence is counterintuitive, finally less persuasive than one which includes the transcendent [Lowentrout, "PsiFi"].

This is quite distinct from recent mocking jibes about cyberpunk or more recently Singularity sf where the transhuman iconography is supposed to evoke a "Rapture of the nerds." If there is a psience fiction rapture, it speaks to our hunger for communication deeper than speech or writing, for a life "when you care, when you love" (as Sturgeon put it in a famous late story title). This seems to bring us back to the Panshins' claim that sf (and, I'd assert, psience fiction especially) confronts the transcendent—not only as a superscientific future condition, but with the human heart, embedded in some way in the grammar of our evolved status as *Homo sapiens*. Lowentrout might be right in saying that

> psi has moved to the background and the tale is not about psi as much as it is about human and natural community in relation to the transcendent. In this, the metaphysical SF psi story is compensatory, our spiritual yearnings driving its development. In times past, it was in relationship with the transcendent that true community, human and natural, was believed to be most perfectly constellated. But unlike the hunter-gatherer societies in which our species evolved, our modern industrial societies have little use for the transcendent, and in them we have little sense of connection with the natural world and only a pallid experience of human community. The metaphysical SF psi story is compensatory [Lowentrout, "PsiFi"].

Ψ

Is this the point of connection between parapsychology and psience fiction? Certainly the current paradigmatic experiments in psi phenomena—remote viewing and presentiment—are precisely utter strangeness seen against a banal familiarity.

As described by scientists engaged with psi, remote viewing is a practice of sitting quietly and then, with mind allowed to roam freely through a possible infinity of time and space, sketching and annotating whatever task-oriented but undisclosed images, scents, feelings come to mind. From this jumble, blinded judges are led to a speculative assessment that the unknown target is, say, a crashed aircraft in a forest beside a river curving to the east before tumbling in a waterfall, possibly in central Africa, with all but one of the crew dead.

Presentiment, the psi favorite *du jour*, is a surge in measures of the autonomic state of the body just *prior* to a random sharp stimulus startling the participant who sits quietly in an ordinary room.

Other protocols equally blend the ordinary and the seemingly unbelievable. Does this require opening an aperture into an irrational unconscious, a kind of transcendence perhaps as difficult as the switch from Newtonian to quantum physics, or Aristotelian logical thinking to higher order math and logics? Possibly so, and a background in reading psience fiction can perhaps help ease entry to this ankle-testing pathway.

Coming from another angle, we might ask: Is there any necessary link between such inducible moments of transcendence and the imagined worlds of psience fiction? Or is psience fiction no better than an escape back into superstition, super-charged adventure, and occasional poignant metaphor, undermined in advance by the comfortable skeptical assertion that *psi is bunk*? Each reader must reach his or her own assessment. But to adapt the concluding words of Mary Roach in an even spookier context: Just as tales of space travel and nuclear weapons foreshadowed a world where those once-derided speculations became daily realities: What the hell—I believe in psi.[4]

Except that this is a misleading way to put it. I don't *believe* in the reality of psi, any more than I *believe* in gravity, or that water is a blend of hydrogen and oxygen; rather, I accept the accumulated *evidence* for its reality. That could change, and if it does I will alter my understanding of the universe. I'm not at all sure, though, that diehard skeptics would do so if the results of future experiments continue to support the reality of psi. In the alternative, would solid proof against psi destroy psience fiction? Maybe so; there's not much demand for phlogiston fiction these days.

Appendix 1
A Brief Guide to Paranormal Research

The vexed history and development of formal psi research started with the rise, in the second half of the 19th century, of spiritualist cults, mediums, ghost-hunters, mind-readers, levitators of tables and mystics alike. In 1882, the Society for Psychical Research was established in Britain at the suggestion of Sir William F. Barrett, and attracted the interest and abilities of significant figures in philosophy and the natural sciences: Henry Sidgwick, professor of moral philosophy at Cambridge (the SPR's first president), W. H. Myers, and others of equal caliber. Their program included studies into telepathy, hypnosis, mediumistic "sensitivity" to alleged messages from the beyond, hauntings, automatic writing, and the like. Its founding program was (as expressed in every issue of the *Journal* of the SPR) to "examine without prejudice or prepossession and in a scientific spirit those faculties of man, real or supposed, which appear to be inexplicable on any generally recognised hypothesis." An American version of the Society, founded by the pragmatist William James, came into being three years later.

By the 1930s, much of this interest in spiritualistic phenomena was ebbing. Some at least of the mediums were accused of fraud, and the tide turned against psychical research as it had first been delineated. But this was also the dawning epoch of the quite new disciplines of psychology and statistics. Dr. J.B. Rhine at Duke University and his colleagues set out to apply these investigative and evaluative techniques. Notable scientists, some of them Nobel laureates, had already looked at possible physics (or paraphysics) of these alleged mind powers, but meanwhile regular science itself was still in uproar with the paradigm-smashing invention of relativity and quantum theory.

So the new studies came to be termed "parapsychology," and conscripted techniques more suited to the lab than to the medium's dimly lit séances. Accounts of spontaneous wonders did not cease, but the emphasis turned to a para-species of Psych 101 bean-counting. Rather promising large initial deviations from chance fairly quickly tumbled under this onslaught of numerical dreariness, but even so the marginal stats did pile up, indicating the action of some mysterious process unknown even to the new physics. These capacities were seen as outside the limits not only of the known senses but of relativistic space and time. Rhine spoke of "the reach of the mind" and hoped to find an explanation of humanity that surpassed the meaningless behaviorism of that benighted epoch, with its staggeringly limited stress on "rats & stats."

After a brief flurry of excitement, and several rather sober volumes by Rhine displaying his findings, parapsychology came under increasingly venomous attack by orthodoxy. The history of spiritistic "woowoo," a derisive term frequently misapplied to psi research, helps explain this hostility. (Oddly, perhaps, simple unlikelihood and paucity of empirical support rarely seem to afflict the standing of university programs into superstring theory that invoke undetected dimensions, and many worlds cosmology, and wormhole studies—all, I hasten to add, in my view admirable attempts to understand aspects of the universe otherwise inexplicable.)

It is easy to see why this would cause the lips of most scientists to curl in derision, even those who go faithfully to church or temple and give some degree of credence to a supernatural realm. Curiously, many routine psi skeptics are also conventional religionists happy to embrace extremely strange claims—virgin births, resurrection from death, elephant-headed gods, guardian angels, the standard panoply we must not mock lest we give grievous offence. Actually, as the Star Gate remote viewers later found in their long government-funded quest for reliable psi applications, adherence by certain highly placed military chiefs to fundamentalist Christianity was also detrimental to the fate of such research, since it was denounced by these individuals as literally the diabolical work of Satan and his minions. (Who else, after all, could perform what amounted to miracles without the sanction of Holy Scripture?)

This tangled history, and the path from it toward experiment and statistics, was described crisply by Dr. Joseph A. Rush (1911–2006), a physicist who worked on the Manhattan Project and later became the first secretary-treasurer of the Federation of American Scientists:

> Parapsychology emerged a century ago out of preoccupation with the question of survival of death. The early research was intermingled with and influenced by spiritist concepts and practices, and it relied heavily upon the "natural history" type of investigation. Now survival appears to be impossible in principle to establish rigorously. Spiritism has all but vanished from professional psi investigations, lingering in some aspects of poltergeists and in a few mediumistic studies [Rush, 1987, pp. 67–8].

It has to be admitted that more than a few of the credentialed teachers and researchers of parapsychology do retain a fondness for philosophical idealism rather than "materialism" or physicalism. Some insist that the universe is built out of primordial Consciousness (whatever that means). A few, as noted earlier, are "panpsychists," as are philosophers Galen Strawson, David Chalmers and Thomas Nagel, who argue that even electrons and quarks must have some minuscule trace of mind or how could it arise from nowhere at our macroscopic scale? (This line of thought is not used, oddly enough, to prove that electrons and protons must be at least a tiny bit wet, because how else could water be wet?)

Some parapsychologists are convinced that life of some kind persists after bodily death, so that mentality is either ported into a new body (reincarnation) or expands, no doubt with a sigh of relief, from the onerous restrictions of the flesh into a supernal realm glimpsed, perhaps, by those who've had Near Death Experiences. This makes it hard to deny that even today psi studies are frequently inflected (or infected) by metaphysical constructs closer to the Spiritism of the 19th century than to the reigning naturalism of the 21st. (Imagined versions of life beyond death have also been a continuing thread in psience fiction, usually invoking new technologies or new theories of mind, and we shall return to this vexed topic in Appendix 2, "Psi and Afterlife in Psience Fiction.")

Ψ

Even so, Dr. Rush was not wrong to note that "The experimental approach has largely replaced the investigation of spontaneous or mediumistic phenomena. Adopted first to establish convincing evidence for psi, it now is directed primarily to elucidating it by exploring its variations and relations of psychological and other variables" (p. 68).

This distinction is sometimes rendered as "proof" versus "process." For most researchers in this discipline, as with the fervent believers in the mediumistic tradition but for better reasons, there is no need any longer to search for *proof*. True, the phenomena are flighty and will not always come when you call; their signature is hardly ever detectable subjectively by psychic agents at the moment psi modifies their experience or indeed the causal linearity of time. Still, suitably rigorous statistics can readily detect, say, an anomalous deviation in a string of guesses when compared to a control sequence. So the basic *reality* of anomalous cognition (and perhaps perturbation, or PK) is established.

One index of this startling assertion is the feeble repudiation that hard-edge skeptics offer after looking closely at the proffered data and analyses. Richard Wiseman, with a doctorate in psychology, studied psi in the lab before stating firmly that psi doesn't exist, because it just plain *can't*. He declared, without embarrassment, that "I agree that by the standards of any other area of science that remote viewing is proven, but begs the question [he meant "raises the question"] do we need higher standards of evidence when we study the paranormal? I think we do" (Wiseman, 2008).

Professor Jessica Utts (in 2017 the 111th president of the American Statistical Association) and Professor Ray Hyman were contracted in 1995 by American Institutes for Research (AIR) to evaluate the Star Gate program. Utts was convinced by the data: "*It is clear to this author that anomalous cognition is possible and has been demonstrated.* This conclusion is not based on belief, but rather on commonly accepted scientific criteria" (my italics; cited from May and Marwaha, 2014, p. 5) Although he remains to this day an adamant skeptic, Hyman stated in his portion of the report that "I tend to agree with Professor Utts that real effects are occurring in these experiments." However, while he couldn't explain them, he was unwilling to take the extra step of accepting that they had anything to do with anomalous psi perception.

Ψ

In short, the rejection of psi as a genuine process in the known spacetime universe is based on what Richard Dawkins calls the fallacious "Argument from Personal Incredulity"—but Dawkins does so only when the fallacy is applied to, for example, the now-accepted theory of evolution by natural selection of heritable variants.

Although the professional organization of academically qualified psi specialists has been an affiliate of the American Association for the Advancement of Science since 1969, its discipline remains the target of shrugs and eye-rolling from most scientists. A cadre of devoted uber-skeptics ensures that this attitude of caustic dismissal penetrates the Internet, press commentary and textbooks alike.

Bear in mind that most people accept as true, or at least truthy, numerous assertions that seem to critics laughable absurd. A Harris poll a few years back reported that only 47 percent of adult Americans who checked off their answers accepted the reality of evolution—meaning that more than half the population still denies the basic lynchpin of scientific biology. Creationists, regarded by skeptics as the genuine wingnuts of evolution denial, comprised 36 percent of respondents. Meanwhile, 42 percent accept the reality of ghosts, 29 percent put their faith in astrology, and a quarter of the population think they'll be reincarnated in another body. (Most of the rest think they'll go to heaven or perhaps eternal punishment in hell.)

Popular acceptance of the paranormal has been clocked at 41 percent or higher.

It turns out that the paranormal believers—probably largely for the wrong reasons—are right to accept the reality of some anomalous phenomena. There is substantial evidence, both anecdotal and from the laboratory,

A Brief Guide to Paranormal Research 199

for some paranormal phenomena. But it takes informed savvy to sort out the trustworthy and interesting from the vast piles of dross.

It's just possible that the next few years will prove in retrospect a key moment in the transition of parapsychology from automatic disdain by most scientists (who have rarely immersed themselves in the data and theoretical arguments of the discipline) to cautious openness. It would be pointless trying to give sufficient details in the limited space of this Appendix to persuade the doubtful, but fortunately it is now easy to find detailed summaries in book form (as well as in many formal papers published in the specialist journals such as *The Journal of Parapsychology*). I'll briefly sketch two current authoritative tomes to suggest the range of the topic and its diverse explorers.

Ψ

By the 1970s, both Soviet and U.S. governments put money into efforts to develop psychic spying. The American program was tasked and funded, under intense and regular scrutiny, year after year for two decades by the Department of Defense as well as CIA, DEA, NSA, NSC, the whole battery of three-letter agencies, not to mention the Secret Service. For the final decade, the scientific wing of this effort was directed by Edwin C. May, a PhD in low energy nuclear physics, who helped shaped the multimillion-dollar Star Gate program until it was shut down in 1995.

May and his Indian psychologist colleague Dr. Sonali Bhatt Marwaha documented all this formerly top secret and classified material for the immense four-volume *The Star Gate Archives*, with data and somewhat redacted operational records, being published by McFarland. This trove might open the way to a truly substantive revaluation of the status of psi phenomena.

May had already co-edited *Anomalous Cognition* (the neutralized term for perceptual psi) and *ESP Wars: East and West*, both released in 2014. In a 2015 two-volume anthology *Extrasensory Perception: Support, Skepticism, and Science* (hereafter *ESP*), he and Marwaha provided expert summaries of the range of recent research into these phenomena at the fringe of orthodox science, along with a dizzying variety of potential explanations.

Simultaneously, a different but convergent endeavor appeared from McFarland, *Parapsychology: A Handbook for the 21st Century* (hereafter *Handbook*), edited principally by Etzel Cardeña, PhD, Thorsen Chair of Psychology at Sweden's University of Lund and Director of its Center for Research on Consciousness and Anomalous Psychology. His main co-editor was Dr. John Palmer, director of research at the Rhine Research Center in North Carolina, and editor of the *Journal of Parapsychology*.

Most of the other contributors to those two stout volumes hold doctorates in related fields; some are professors. Each book has a detailed early chapter on suitable methods for psi research and attempted replications by Professor Jessica Utts. This is not an assembly of cartoon mall-front fortune tellers peering at grubby Tarot cards and crystal balls.

That's not to say spirituality doesn't make itself known even in up to date surveys like these, especially in Cardeña, et al. Four chapters in *Handbook* explicitly confront purported evidence for life beyond death: mediumship, reincarnation field studies, ghosts and poltergeists (often seen as anomalous perturbation or angst-driven psychokinesis rather than cavorting dead people) and, arguably the least interesting of all, supposed "electronic voice phenomena" (EVP) where random sounds seem to form a few second of postmortem or maybe demonic utterance. There's a word for at least most of this: Pareidolia, the trick of the mind that lets the devout see the Blessed Virgin scorched into a slice of toast or a pair of underpants. The EVP chapter writers are candid enough to report that "confidence in one's interpretation does not usually reflect that other people will hear the same thing."

On the other hand, a quite impressive *Handbook* attempt is made by Edward F. Kelly, PhD, a research professor in the Division of Perceptual Studies at the University of Virginia, to demonstrate that the mind can't possibly be identical with the embodied brain. It must be invisible and neither material nor energetic, and able to survive death. How, then, can it couple to the brain and other organs? By psi, perhaps? We could call this doctrine *promissory immaterialism*. It's an opinion seldom heard from neuroscientists, other than Christof Koch, but as noted above it has increasing cachet among panpsychist philosophers such as Galen Strawson and David Chalmers. It is a background beat in the understanding of some parapsychologists such as Stephan Schwartz and Charles Tart, who consider that Mind not only permeates the universe but is its source and substrate—a view also at the basis of Hindu theology. Others regard that notion as a category mistake, a term from Gilbert Ryle's most famous book *The Ghost in the Machine* (or in this case the machine in the ghost).

$$\Psi$$

What surely would have appealed strongly to John W. Campbell are the substantial and various treatments of physicalist theories of psi: how does it work, why does it often *fail* to work reliably, how might the tremendously powerful "ways of knowing" of advanced science accommodate their equations to intrusions that seem both ridiculous and impossible?

But isn't quantum theory exactly like that, some ask. What about *entanglement*, when the states of correlated twin particles remain suspended in double possibility until one happens to interact with the rest of the universe and decoheres into definiteness, and the other particle instantly does the same *even if it's a light-year distant*? Isn't that a key to faster-than-light knowledge of the future and of distant events outside the light cone? No, unfortunately. The correlations are real, but there's no known way to modulate them into a signal that conveys useful information. Indeed, quantum theory sternly refuses to allow this.

What, then? In *ESP* Vol. 2, Bernard Carr, professor of mathematics and astronomy at the University of London, invokes hyperspatial dimensions like those posited by string theory. Edwin May, and Joseph Depp, PhD, ponder wormholes through spacetime like those postulated by general relativity. The late Richard Shoup, a computer science PhD, took an even stranger step sideways. Shoup more or less invented programmable logic and reconfigurable hardware and was probably the only psi researcher to win an Emmy and an Oscar for his technical work. At the Boundary Institute, he mapped the outlines of a research program that replaces causality with correlation, based on the time reversibility of fundamental physics. Daniel P. Sheehan, PhD, professor of physics at the University of San Diego, argues for a link between precognition and just such retrocausation.

Other arcana from German researcher Walter von Lucadou, PhD, senior lecturer at the Furtwangen Technical University, presents a model of pragmatic information, which claims that psi *must* get weaker or at least more dispersed as it gets closer to actually *working usefully*. Why? Because the universe can't abide time loops and paradoxes. It's an amusing notion, but seems to come entirely unstuck when tested against the vivid empirical reports from Joseph McMoneagle (in *ESP* Vol. 1), the best known and perhaps most successful remote viewer from the Star Gate program. His ability to sketch and describe distant military and other objectives with high (but not perfect) accuracy extended satisfactorily into the future.

Is there any way to capture the dynamics of such mystery processes? Edwin May's discussion of entropy as a key to precognition hovers, for the non-scientist, on the brink of intelligibility and maybe beyond the edge. So also does Dutch professor Dick J. Bierman's "Consciousness-Induced Restoration of Time Symmetry." Still, while it is too soon to be sure, something like one of these models must account for psi, because simple fraud and stupidity surely don't suffice. John Campbell was surely wrong in supposing, all those decades ago, that he already had the basic rules of psi under his thumb. But sooner or later, somebody will. Unless the entire topic is, after all, a tragic mistake.

Ψ

Remarkably, both books contain a substantial chapter by critics of the evidence for psi, and perhaps of the very idea. May and Marwaha's chosen debunkers are Dr. Eric Jan Wagenmakers, professor at the Psychological Methods Unit at the University of Amsterdam, and four associates. Using

Bayesian statistics rather than traditional old-school frequentist analysis, Wagenmakers builds in his pre-existing judgments or biases, gained from the complex experience of all other assessments he's ever done, and modifies these only if necessary by the deviations from chance found in the latest experiment.

A Bayesian is forbidden to have a prior likelihood estimate of zero—after all, anything *might* turn out to be true, and perhaps we all inhabit a prankish Second Life–type virtual cosmos—but we are allowed to maintain a probability of 0.00000000001 or lower that the world is actually flat, or that psi is real. Given such a starting point, it can still be impossible ever to change your mind, no matter how strong and consistent the evidence (and psi's is rarely either strong nor stable, generally just cumulatively anomalous in the long run). Wagenmakers' interest in what he feels in his bones to be absurd nonsense is frankly this: "if our standard scientific methods allow us to prove the impossible" (as parapsychology does) "then these methods are surely up for revision" (Wagenmakers et al., p. 156). This used to be known as burning down the village to save it. Since he does not fear this outcome, then we are left with psi as "a control condition for science, an unwitting jester in the court of academia."

Handbook's skeptic is Dr. Douglas M. Stokes, a mathematical psychologist and former associate editor of the *Journal of Parapsychology*, who has become disenchanted with the whole domain of inquiry, convinced that there is more hanky-panky possible than meets even the astute eye. It is a position seemingly supported by recent convulsions in the social sciences, where a scandal erupted over Questionable Research Practices (QRPs) in many disciplines, most of them until now regarded as securely authoritative. There are numerous ways to bias results to achieve a desired outcome, many of them unconscious or apparently non-toxic: optional stopping when the numbers look good, filing away experiments with poor outcomes so only the winners get published, a dozen more.

In 2016, a paper by anomalies researchers Dick J. Bierman, James P. Spottiswoode and Aron Bijl, "Testing for Questionable Research Practices in a Meta-Analysis: An Example from Experimental Parapsychology," applied these QRPs directly to psi research (although without any experimental corroboration that questionable moves actually *had* influenced the reported results). The revised assessment dragged down the significance levels from amazingly impressive six sigma scores to, say, a chance probability of 0.003.

That still remains moderately strong evidence for an anomaly—since getting this result by sheer fluke will happen only three times in a thousand repetitions. And it depends on assuming that parapsychologists are guilty in the same degree as their worst mainstream colleagues. Scornful doubters will proclaim they are *more* guilty, on the skeptical assumption that psi just plain

A Brief Guide to Paranormal Research 203

doesn't exist. Fortunately, academic psi journals were among the first to insist on publishing failed experiments alongside successful ones—a policy that contrasts favorably to the unwritten but powerful rule in the loftiest major journals like *Nature* and *Science* that only startling new results are worth the paper or electrons they're printed on.

Ψ

There are other startling results that call for an expanded science or psience—rather as the precession of Mercury needed relativity to explain it, and what was called the atomic ultraviolet catastrophe required quantum theory to explain why it didn't happen. *Handbook* has chapters by Dr. Dean Radin on presentiment or presponse, the detectable physiological jolt in skin conductivity or blood pressure or subtle brain states *prior to* a random shock. (It might be that presentiment effects, experienced as barely conscious intuitions, save us from the even greater road carnage one might expect from driving on a murderous freeway.)

Psychotherapist Dr. James Carpenter explains his model of First Sight, which posits an even more general on-going unconscious psi responsiveness to the world, guiding us away from danger and toward benefits. No matter how slight this influence is, it gives evolutionary selection a lever to fix psi in our genome—although perhaps, due to contests between rivals, at the low level we observe. That is an argument, among others, investigated by Professor Richard Broughton in the *Handbook*.

A chapter on the experience of remote viewing in *ESP* Vol. 1 is notably persuasive: a transcript account of military remote viewer Joe McMoneagle's actual 1979 search through his present and future, tasked by the National Security Council, for details of a hidden Soviet weapons program. Step by slow, double-blinded step, he "walked" his point of view or hallucinated spectral body through a great land-locked concrete building in what proved to be the incomplete and unknown Soviet TK089 Typhoon Submarine base. Absurdly, it seemed, this was located some distance from the sea. The Typhoon (or Shark) was a huge nuclear powered ballistic missile sub.

McMoneagle described it in some detail, but CIA's chief Russian desk office, Robert Gates, dismissed the report as "Total Fantasy!" McMoneagle got word of this and wrote back: "The fantasy will be launched in approximately 112 days! J.M.!" Admiral Stewart carried this back to the NRC, and a search of the site was conducted 114 days later. A huge channel had been cut to the sea, and a Typhoon sat at the harbor's dock. Gates's response? "Lucky guess!"

Years later, in 1995, Robert Gates was involved in closing down the Star Gate program. He publicly declared on television, with a straight face, that nothing useful had ever come from its decades of research and operational advice. In 2006, he became the 22nd U.S. Secretary of Defense, remaining in that post until 2011. Today Joe McMoneagle is working with Dr. May's Laboratories for Fundamental Research, and still doing surprisingly effective remote viewing. McMoneagle comments:

> Director of the CIA Robert Gates said on *Nightline*: "In no case did the Army program provide any information of value as standalone information." That was a bald faced lie. When they cancelled the program, for whatever reason, the public reason given was that "It simply didn't work, or provide any critical information of value." That too was a bald faced lie. They tasked us for 19 years, close to 65% of what we did was for them, we were lucky to get any feedback directly from the tasker (CIA). We usually read about it in the papers later, or got it through another agency. My own Legion of Merit [decoration] was for providing more than 160 elements of critical intelligence unavailable from any other source, at least according to the Secretary of the Army [personal communication, January 24, 2017].

The January 2017 release of 13 million pages of declassified CIA documents, one basis for May and Marhawa's large archival project, should have a serious impact on honest skeptics, but Joe McMoneagle went on to note the limitations of even this trove: "Hundreds and thousands of pages supposedly have been released by the CIA, et al., on Star Gate. But not all the work has been declassified. An example would be my drawings on the XM-1 Tank, or my cutaways on the Typhoon Submarine. I doubt these will ever see the light of day, or the final reports. Sadly, because they say they've released things doesn't mean they have. These final reports in many cases are still classified."

Some taskings he is permitted to talk in detail about "are out there. Typhoon submarine; bugs in the new addition at the Embassy in Moscow; the spy inside the CIA; Live reporting of the incident at Desert One; monitoring of the Iranian hostages; General Dozier; Colonel Higgins; etc."

Are these successes just happenchance? No, they were too detailed and too useful, when other intel had proved unproductive. Did the remote viewers guess or embroider on the basis of subtle clues in the tasking material? No:

> Tasking arriving at our unit was thoroughly masked so our boss couldn't associate it with anything going on in the world. We were tasked with sealed envelopes. We produced transcripts and extracted information from them, forwarding them back to the tasking agency. And the feedback in some cases said something like: If it weren't for your work and effort we would not have been able to accomplish the mission. Or, You nailed the information we were looking for, even though we didn't specifically task it. All done double blind of course, at least while I was assigned to the unit. I get so tired of all the quibbling over whether or not it was real. Of course it was real. In some cases, it was the only information that resulted in rescuing a hostage who could have been killed or tortured to death; on other occasions, they blatantly ignored information on hostages being tortured, because someone in the loop decided they didn't like

the idea of acting on psi (until they found out four months later it was all correct and might have saved the poor guy who died a horrible death).

How did this military RV protocol work? Were there sophisticated content analytics applied to the scraps of data and sketches provided double-blind by the viewer? Were subtle allowances made for biases in what they imagined they saw? No, the process appears to have been much more direct and subjectively assessed than that:

> I can't speak for anyone else, only myself. What I did was two things. First, I quickly understood that no one outside the [Star Gate military] unit understood RV. Given that was the truth, there was no way they could ask a competent question of a remote viewer. So, I had to find a way to get to the answers they wanted without having to teach them one on one. So, I modified my tasking and never told anyone. Regardless of what their tasking might be, the last thing I said to myself before producing any material on any task was—"Let me give them whatever is going to make them happy and address their problem." This seemed to work extremely well. To the point it sometimes frightened the client.
>
> Second; when you can't get feedback, then you pal up with someone and do practice targets randomly chosen from a huge target file. Our target file was any page with a photograph on it randomly chosen out of a random *National Geographic*. If there was an explanatory paragraph to go with it, all the better. Then I and one other viewer in the group challenged one another to do better and better with the practice targets. I did at least three a day, 7 days a week, for nearly nineteen years. When I retired from the Army, I had lots of business through my company *Intuitive Intelligence Applications, Inc.*, and my customers gave me full feedback no matter what, or I stopped working for them. I still do one target a day for practice or I can't maintain my expertise.
>
> I have no idea how anyone would manage to maintain their ability otherwise [January 25, 2017].

And who first devised this remarkable and remarkably strange activity? Psience fiction? In a way, perhaps. Some like to cite an experiment (according to Herodotus, in Chapter 47, section 2 and beyond, of Book I of his 5th century BCE *The Histories*) conducted by King Croesus, who feared an impending onslaught from the Persian Cyrus. In keeping with the beliefs of the day, he wanted a sibyl's oracular prophecy to guide his response to this threat, and sent out messengers hither and yon with a preliminary test. The oracle of Delphi replied with a poetic hexameter verse describing an extremely unlikely event happening far away: Croesus cutting up a tortoise and a lamb and boiled their meat in a brass cauldron covered with a brass lid. The King was astonished by the accuracy of this exercise in remote viewing. Alas, his empire came to a sticky end when he trusted a further utterance from the sibyl telling him that if he undertook war with Persia he would destroy a mighty empire. This did come about, but the empire was his own. Today's versions of the RV protocol tries to avoid such ambiguities, but the impressionistic diagrams and notes by the chosen viewer often remain somewhat open to interpretation—which explains some of the misfires.

Ψ

McMoneagle provides some background that is generally acknowledged by parapsychologists:

> In my opinion the creator and father of remote viewing was René Warcollier (1881-1962), a French chemical engineer and early psychical researcher. He did what looks exactly like the protocol in RV for more than 30 years with a single subject, his daughter, published and presented his findings at the Sorbonne, Paris, in 1946. Ingo Swann certainly helped in the original development of the protocol at SRI–International, but he did very little, if any, Intelligence Collection other than targeting our stuff as proof in principle. Ingo also helped to develop the concept that you didn't need to have an "outbounder" at a target, you could use GPS Coordinates instead, and he participated in most of the science developed within the first eleven to twelve years.
>
> GPS Coordinates were something the CIA balked at within four months of our using them at Fort Meade. They accused excellent remote viewers of having photographic memories, which was silly. My memory is pretty good, but certainly not that good. I suggested going directly to sealed envelopes, and when the tasking agent arrived having left the envelope on his desk one morning, I suggested he just say 'target the envelope on my desk' and put it in an envelope. That worked just as well, which turned me on to pursuing a paper regarding INTENTION, ATTENTION, AND EXPECTATION FOR OUTCOME.
>
> I've been to Russia and spoken with the movers and shakers inside the Russian Unit, and they do not do RV, and if they claim to, it's nothing like we do.
>
> Hal Puthoff and Russell Targ and Edwin May did the world a great favor; they put the full energy of science behind remote viewing, and as far as I'm concerned proved RV to be about as real as rain, dirt, or the planet we ride around on. So, one could call them the fathers of the science behind RV. Had they not done it, all the millions of dollars that SRI–I and SAIC shared with a great number of researchers out there would never have happened, I wouldn't be here now, and thousands on the Internet who really haven't a clue what RV is would never have heard of it [January 25, 2017].

Except, in a way, through the varied visions of psi, both military and civilian, presented by psience fiction.

Appendix 2
Psi and Afterlife in Psience Fiction

The concept of a psychic or spiritual life for humans beyond individual death and physical corruption is very old and very widespread—as old and universal, perhaps, as dreaming. Afterlife is often analyzed by scientific naturalists as a consolatory fabrication devised out of grief and wishful thinking. Beyond death, then, is an imaginary realm where loved ones persist somehow as if they had traveled to a land beyond the hill or shore. In this place, the evident injustices of mortal life are redeemed and set right, with punishment for the wicked and joyful rewards for the virtuous—or at least those who have successfully petitioned God, gods, or other supernatural forces of immense power and a capacity to intervene on our behalf.

Despite its evident gratifications, it is arguable that the wellspring of this idea is the real, confusing experience of half-remembered dreams.[1] When we sleep, our drowsing minds mingle memory and fancy, placing us or our viewpoint surrogate inside a kind of shifting, surreal virtual reality where time loses its implacable dominion, where the dead walk among us, where strange chimeras are built from fragments of creatures, people, places, motivations and feelings carried over from waking empirical life.

It is easy to see how such imagined worlds, vivid and more various than humdrum narrow reality, might have enthralled our ancient ancestors, undistracted by reading, movies, television, easy travel or frequent visitors. Certainly we know that hunter-gatherers were given to punishing the living for slights or crimes experienced only in dream, in much the way diseases and accidents were widely blamed on sorcery and ill intent.

But even if these are the sources of such widespread and poignant beliefs, are they necessarily untrue for that reason? Parapsychology suggests that one's intentions *might* act on others without any conventional medium of influence, that thoughts might be intercepted even if unspoken. Is it possible that fantasies and reports of life beyond life also offer us glimpses of a reality that scientific cultures dismiss due to their elusiveness and similarity to delusion and psychotic or protective self-deceit?

One interesting intersection between such old beliefs and the scientific *Weltanschauung* is the form of storytelling we have been calling psience fiction, and more generally science fiction. For sf, the known world is all too narrow and restricted. If its narratives freely conjure story-devices at odds with what scientific investigation tells us about reality—time machines to the past, vehicles or messages faster than light, extravagant psi powers—still their deployment remains faithful to the *impulse and methods* of science. Coherence and plausibility have to be retained. A story might violate currently known physics, but it has to provide some sort of quasi-scientific rationale for doing so, or at least play with the net up during its game (as scientist-writer Gregory Benford has put it).

So how does this newest of narrative methods deal with one of the most ancient human concerns: the hope for a life of some different and even transcendent kind following our all-too-familiar life on earth?

Oddly, the vast and authoritative online *Science Fiction Encyclopedia* (since 2011, and expanding all the time), edited by John Clute, David Langford, Peter Nicholls (emeritus) and Graham Sleight (managing), contains no thematic entry for AFTERLIFE. This topic does receive an entry, along with an even longer treatment of POSTHUMOUS FANTASY, as well as REINCARNATION, in *The Encyclopedia of Fantasy* (1999), edited by John Clute and John Grant. The distinction reflects the way commercial imaginative fiction split into the twin modes of *science fiction* and *fantasy*. The former maintains a clearer vocation to realism (although often of an inflated kind), while fantasy's adherence is to something more Gothic and both gaudy and shadowy.

So the metaphors of sf and psience fiction are intended, by and large, to be taken literally, at least in the imagined reality of the story, alongside the descriptions of psychic abilities and persistence beyond mortal death. Those of fantasy remain somewhat allegorical, parabolic, dreamlike. So it is understandable that the larger part of fanciful fiction dealing with an afterlife is couched in the older forms of frank fantasy, where angels, fairies, ghosts, haunts, heavens, hells and gods are part of the familiar landscape, not an intrusion to be rationalized and treated theoretically.

Yet since it is the ambition of parapsychology to deal with topics such as telepathy, precognition and perhaps even life after death on the basis that they are as real as Twitter and Google's offerings, sf's and especially psience fiction's approach to afterlife is probably more salient than fantasy's, even if the pickings are thinner on the ground and perhaps less emotionally appealing or moving.

Ψ

In Robert A. Heinlein's early novel, *Beyond This Horizon* (serialized 1942), the quest by jaded protagonist, Hamilton Felix, for proof of a life beyond death is all that makes his future utopia bearable. When that proof is found, ambiguously, it is more of the same—survival recycled via reincarnation. An essence of the dead informs new infants, although memory of a former life is swiftly lost. It is a position interestingly congruent with the claims of the late Professor Ian Stevenson at the University of Virginia, in his series of detailed studies into cases compatible with reincarnation (*Twenty Cases Suggestive of Reincarnation*, 1980, *Children Who Remember Previous Lives*, 2001, etc.), perhaps the more interesting because Heinlein's serial first appeared more than 75 years ago.

Paradoxically, perhaps, this hard-headed engineer had a keen interest in Theosophical and magical doctrines, and elements of both emerge in many of his stories. In the late novel *I Will Fear No Evil* (1970), its biblical title speaks directly to the theme, in which a very old white plutocrat's brain is transplanted into the healthy but brain-damaged body of a beautiful, young and possibly black woman—whose consciousness somehow remains intact. They conduct an internal dialogue throughout the novel, and while it is not clear until the end if this is a kind of hallucination or a form of "somatic memory," it seems clear that each has a soul that finally passes into a different kind of realm.

That realm seems to be the sort of Ur-state of a partitioned divinity suggested at a key moment in *Beyond This Horizon*, when ill, unconscious Hamilton momentarily merges with a game-playing Mind that seems to shift from one personality to another, and even change games (and universes) at whim.

> It was pleasant to be dead.... The next time he would not choose to be a mathematician. Dull, tasteless stuff, mathematics—quite likely to give the game away before it was played out. No fun in the game if you knew the outcome.... It was always like this on first waking up. It was always a little hard to remember which position Himself had played, forgetting that he had played all of the parts. Well, that was the game; it was the only game in town, and there was nothing else to do. Could he help it if the game was crooked? Even if he had made it up and played all the parts.
> But he would think up another game the next time [1981, pp. 152–53].

Obviously this is by no means a conventional religious or mystical premise, not even for an Eastern faith. It is more like the kind of solipsistic *faux*-religious sf apparatus later developed by Heinlein's fellow pulp writer for *Astounding Science Fiction*, L. Ron Hubbard. From such an uncompromisingly Cartesian dualist perspective, the only thing that indisputably exists is an observing, constructing mind, but that mind is not identifiable with any one of us. It is not so much that there is an afterlife following this one; rather, our mortal lives are small fictions or roles played out within innumerable "fictons," or alternative realities, as suggested slyly in *The Number of the Beast*—(1980), where Heinlein is himself, as Author, the Beast under various guises.

Ψ

None of this implies that science fiction writers have avoided the marginally psience fictional theme of an afterlife, although it is true that sf is typically reductive, materialist and atheist in orientation. If humans attain transcendence, as in the closing movements of Sir Arthur Clarke's *Childhood's End* (1953) and the Sturgeon novels discussed previously, it is usually via incorporation into a kind of cosmic Overmind that is not so much divine as immanent in the spacetime structure of the universe, or deeper still. The format of earthly minds—of brains in bodily action, complete with memories and senses—is somehow written into a more permanent and subtle form of matter or energy field.

This resembles the iterated (if reductionist) fate of the deathless citizens of Diaspar, in Clarke's *The City and the Stars* (1956). More than a billion years hence, the citizens of this last and greatest city live for a millennium and then are dissolved back into computer-stored files, to be embodied again many millennia later, memories returning in maturity. And in Clarke's *2001*, this transformation is reported as the history of the first denizens of the galaxy, and by implication as our own destiny:

> The first explorers of Earth had long since come to the limits of flesh and blood; as soon as their machines were better than their bodies, it was time to move.... In their ceaseless experimenting, they had learned to store knowledge in the structure of space itself, and to preserve their thoughts for eternity in frozen lattices of light. They could become creatures of radiation, free at last from the tyranny of matter.... Into pure energy, therefore, they presently transformed themselves.... Now they were lords of the galaxy, and beyond the reach of time. They could rove at will among the stars, and sink like a subtle mist through the very interstices of space [1968, p. 184].

This is not so much an afterlife as the emergence of a butterfly from a pupa. In Bob Shaw's *The Palace of Eternity* (1969), souls are *egons*, immortal self-sustaining extraterrestrial patterns of energy that attain awareness via rapport with material, planet-bound creatures. "As the physical host grows and matures, his central nervous system becomes increasingly complex." This development is matched by the egon, which is set free when the host dies. "Equipped with a identity, a highly complex pattern of self-sustaining energy, it is reborn to its heritage of endless life" (1969, p. 132)

But if something like this were true, why would that fact give us humans comfort? Because, Shaw proposes, "as far as the host is concerned, death is merely the doorway to this new life—because he is the egon." The same idea informed veteran sf author Clifford D. Simak's *Time and Again*

(1951); Brian Stableford notes that his "alien symbionts which infest all living things are obviously analogous to souls" (Clute and Nicholls, 1993, p. 1002).

This is a curious inversion of the old legend of the vampire, which sucks the life out of the warm living. (Vampires, of course, are neither dead, "passed across," nor resurrected, but uncannily *undead*.) Indeed, Colin Wilson's garbled but intriguing quasi-sf novels adapted this theme: explicitly as predators on human "life-fields," in *The Space Vampires* (1976), but more metaphorically in *The Mind Parasites* (1967), which drew upon the horror mythos of H. P. Lovecraft to construct an eerie secret history of humans as prey to disembodied creatures.

In a long sequence of novels about the immense Riverworld where all postmortem humanity awakens (*To Your Scattered Bodies Go*, 1971, *The Fabulous Riverboat*, 1971, *The Dark Design*, 1977, *The Magic Labyrinth*, 1980, *Gods of the Riverworld*, 1983, etc.), Philip José Farmer proposed, like Simak and Shaw, that we are born lacking immortal souls. In his cosmos, compassionate aliens will someday construct a sort of plug-in spiritual module or "watham" for the technologically resurrected dead. These henceforth *will* enjoy the kind of deathless preternatural life once ascribed by Catholic theologians to unbaptized infants in Limbo. Alas, fallen humankind makes a hash of these plans. It is the same bleak image of human afterlife as Poul Anderson's "The Martyr," 1960, where it turns out that humans, unlike aliens, simply don't *have* immortal souls.

A somewhat similar notion plays out in several novels by Spider Robinson (*Time Pressure*, 1987; *Lifehouse*, 1997), in which time travelers from the future create a device in the past that captures and sustains the memories of the dying, salvaging all of us from mortal extinction. This general idea was advanced quite seriously by physics professor Frank Tipler of Tulane University. In *The Physics of Immortality* (1995), he argued in prodigious detail, complete with mathematics and game-theory equations, that a closed universe will evolve toward a Teilhardian Omega Point god state that will recreate or resurrect every person who has ever lived, and indeed every person who *might* ever *have* lived, to enjoy an endless paradise packed exponentially into the closing fractions of a second of the extinguishing Big Crunch cosmos.

It is a breathtaking perspective, and not particularly credible (even less so now that we know the cosmos seems doomed to accelerate and cool endlessly rather than collapse in a reverse of the Big Bang's fires), but has been adopted by a number of sf writers such as Frederik Pohl in his Eschaton trilogy (*The Other End of Time*, 1996, *The Siege of Eternity*, 1997, *The Far Shore of Time*, 2000). Again, while these approaches to an afterlife are audacious and based to some extent on recent physics, they do not at all evoke the traditional images provided by anthropology, theology or mysticism.

Ψ

Neither does a more plausible middle-term possibility discussed in both sf and futurist studies: the uploaded personality. If a mind is just the brainy body in action, responding to the world, to other people and to its own memories and conceits, then in principle it might be feasible to make a one-to-one mapping between each working element of the brain and a more durable machine substrate.[2] Greg Egan explored this kind of notion in a number of stories ("Learning to Be Me," 1990, *Permutation City*, 1994, *Diaspora*, 1997), as did Fred Pohl (the *Gateway* sequence, 1977, 1980, 1984, 1987, 2004), John Varley ("Overdrawn at the Memory Bank," 1976), Charles Stross (*Accelerando*, 2005) and many other sf writers. In some of Egan's stories, future citizens have a "jewel" (or dual) implanted hygienically in their brain at an early age. Each neurological action is detected and copied into the vast storage of the jewel. Finally the internal structure of the jewel has become an exact copy of the brain it shares, running in parallel to its organic original, and the vulnerable brain itself is disposed of, like the vestigial vermiforme appendix.

The upload option, creepy as it must seem at first, promises a kind of non-psi immortality, since it amounts to recording a regular backup of your mind-state that can be reinstated if your present instantiation dies. The new you would lack your most recent memories, especially if you died violently. "She started quivering again. The person who had written that final paragraph," reflects the revived or downloaded backup of a character in Ken MacLeod's *Newton's Wake* (2004), "was a person different from herself.... Her other self had been changed ... by some experience other than approaching death, in some way that her present self could not understand" (p. 211).

Of course that sort of machine-mediated afterlife is entirely different from the kinds imagined by religions or other supernaturalist doctrines. What's more, many critics (I am one, and MacLeod too, it seems) refuse to accept that a restored copy *is* the lost person; rather, he or she is just another person sharing the original's memories and concerns.[3] At the minimal extreme, an upload is used in sf as a sort of convenient talking book or oracle, the personality of the dead stored indefinitely and accessible, even if somewhat mechanical in tone, as in Cordwainer Smith's "Alpha Ralpha Boulevard" (1961).

Meanwhile, interestingly, the established Christian churches are less exercised than one might expect on topics such as imminent machine intelligence and perhaps machine consciousness. In a remarkable address in Rome in December 2004, the president of the Pontifical Council for Social Communications, Archbishop John Foley, declared that the evolution of technology now raises the question whether "it is about humanizing the machine or about transforming man into something inhuman." The president of the Pontifical Council for Culture, Cardinal Paul Poupard, added that while "the machine seems to be the negation of man and robotics, the annulment of the spiritual dimension" (a fear consonant with Philip K. Dick's sf and psience fiction obsessions), still "God has given man intelligence, which has enabled him to produce ever-more sophisticated machines, and has left him free to make his choices. We are the ones who create our technological reality" and "confront the evolution of a new dimension in which human intelligence is united to artificial intelligence."[4]

That is a remarkable admission, and one that eventually might help bridge the gap between ancient dreams of an unprovable afterlife, and techno-dreams of afterlife mediated by the kinds of mechanisms imagined so far mostly in science fiction. Especially if a psionic machine is developed capable of reading and storing an organic human's mental contents, somewhat like the half-life in Philip Dick's *Ubik* but preferably far more stable.

Ψ

What if the whole world of experience is a sort of elaborate and deceitful simulation, as in the movie trilogy begun in 1999 with *The Matrix,* and the little-known Daniel Galouye novel *Simulacron-3* (1964) that preceded those movies by decades and was filmed by Joseph Rusnak, with Armin Mueller-Stahl, as *The Thirteenth Floor* in 1999? This is even more explicit in Christopher Nolan's *Inception* (2010), with both shared and solipsistic dream worlds. It is not inconceivable. Already we have quite lifelike and quasi-artificially intelligent toy worlds in computer game form. If our own civilization continues uninterrupted, growing in computational prowess and raw power, our posthuman descendants might choose to run simulations of our present epoch.[5] Suppose they make very many of these, indistinguishable in their fine grain (especially to the simulated inhabitants) from the original world? Well, then, considerations of probability suggest that our own world is most likely to be one of these simulations (see several papers by the philosopher Nick Bostrom,[6] now famous for his study of the prospect of superhuman artificial intelligence).[7]

In such a nested cosmos, it is possible that death can close down not only each individual human consciousness but the entire world—not in a redemptive religious Rapture, or cosmic calamity billions of years hence, but at any moment, if the programmer running the simulation or emulation grows bored and deletes it. (See economist Robin Hanson's remarkable non-fiction thought experiment *The Age of Em: Work, Love and Life when Robots Rule the Earth,* 2016.) On the other hand, a kind of afterlife might be found if the same personality templates or algorithms are reused in various quite different simulations, or if particularly delightful or wicked sims or ems are recorded as art works or object lessons, to be archived or replayed at will. Of course, this prospect is nothing at all like the afterlife as reported by spiritualists and explorers of Near Death Experience, but those, too, have been the source of interesting sf, most notably Connie Willis's award-winning novel

Passage (2001), which offers a plausible but debunking evolutionary explanation for NDE as complex hallucination.

Ψ

In short, the theme of the afterlife has been taken by sf writers down many odd and hitherto-unthought roads, as well as many that are well trodden.[8] An entire anthology of exemplary sf stories is *Afterlives* (1986), edited by Pamela Sargent and Ian Watson. The editors offer "aliens providing an afterlife and advanced human technology producing artificial resurrection ... the 'timeless moment,' as well as vicious warfare in the godless heavens ... the passage between life and death ... afterlives that are bizarre, happy, obsolete, inverted, frightening, tragic and comic" (xvii).

That is a neat summary of sf's conceptual approach to afterlife: all and anything, except, generally, the solemn and sacerdotal. Sargent and Watson's useful gathering does neglect a little-known but unnerving "Paratime" oddity by H. Beam Piper, "Last Enemy" (1950). In a parallel universe, reincarnation continues to segregate people on class and ideological lines. Also omitted is one of the most famous tales of a zombified afterlife, SFWA Grand Master Robert Silverberg's "Born with the Dead" (1974), a literate and chilly portrait of a man obsessed by his dead Eurydicean wife, revived to a passionless or at least incomprehensible state by future medicine; a kind of Orpheus, he chooses finally to join her in living death, and by inevitable bitter irony shares her cool, unattached condition.[9]

This is an ancient dread of life beyond death: that it will be calculating, distant, unemotional, or at best menacing and cruel. It is the vision of death at the heart of Ursula K. Le Guin's *The Farthest Shore* (1991), where death is a remote silent realm akin to an Egyptian frieze, or the more recent aching, awful vision of Philip Pullman in *The Amber Spyglass* (2000), the concluding volume of his immensely rich Miltonic trilogy, *His Dark Materials*. These two are fantasies rather than science or psience fiction proper, but their methods follow in most ways the prescription of sf: careful creation of variant worlds with their own lawful rules not to be broken by caprice. It is significant that Pullman's sequence describes as a liberation the literal withering and death of God (or god, or Gnostic Yahweh), and the freeing of the trapped dead—each of us, it turns out, has a glum personal Death that accompanies us in life, and beyond the grave—in a sort of redemptive evaporation "into the night, the starlight, the air ... gone, leaving behind ... a vivid little burst of happiness" (2000, p. 364).

Ψ

Then there are liminal states, ontological disruptions at the margins of death and life, a favorite haunt of Philip K. Dick. In *Ubik*, his characters are frozen in a cryonic half-life after a fatal explosion, their minds gradually ebbing, leaking into each another, fabricating simulacra of realities into which bizarre and blackly comic irrealities intrude. Is this the shape of an afterlife we might yet attain if cryonic suspension is developed further, so that many people wait, after clinical death, in a sort of chilly ambulance to future resurrection?[10]

Philip Dick's recurrent obsession was a standoff, as he saw it, between the human and the android: between, that is, the poignant existential reality of *qualia* and sensitivity, versus a cold pretense in the form of machines that nag, pester, advise, loom, threaten, displace the human. In some stories, a protagonist learns with dismay that he *is* a machine, or even a programmed bomb aimed at real humans. In *Eye in the Sky* (1957), the world is an illusion, or series of illusions, created by several characters at odds with each other at the moment of their death. In the movie *Bladerunner* (1982), loosely based on Dick's fine and complex dystopia *Do Androids Dream of Electric Sheep* (1968), an android-killing operative faces artificial people as complex and passionate as himself. Indeed, there are hints (in the movie, though not in the novel) that he is himself an android loaded with fake memories. If an afterlife is the blessed repose for souls graduating from this Vale of Tears, may an android dream of Ecumenical Sleep?

Might any artificial consciousness—assuming (as a large number of sf works do) that such beings will someday be constructed—be deemed to have a "soul" capable of salvation and endurance beyond this life? The question might be sophistical, as most unbelievers maintain. Or it might be that a sufficiently complex mental structure, whether organic, silicon or quantum, generates some kind of organized field that might persist after its substrate's death. This notion has been advanced (as noted above) by Professor Johnjoe McFadden, and by parapsychologically informed biologists and physicists such as Rupert Sheldrake and the late Evan Harris Walker. It was a Campbellian psience fictional staple in the fifties and sixties, and influential in other major sf magazines as well, such as *Galaxy,* where Robert Sheckley's *Immortality, Inc* (1959) was first serialized in 1958–9 as "Time Killer." Sheckley's method of immortality amounted to hijacking the bodies of others, a putative form of vampiric afterlife that some spiritualists dub "drop-ins" and is displayed in the case of Victor in Dan Morgan's Sixth Perception sequence (Chapter 28, above).

That trope is used on a very large scale in Peter F. Hamilton's "new space opera" trilogy, "Night's Dawn" (*The Reality Dysfunction,* 1996, *The Neutronium Alchemist,* 1998, *The Naked God,* 2000), where dead souls erupt to infest the living across many stellar systems and thousands of pages. The legacy of shamans, spirit seers, channelers, Raudive electronic voice interpreters, NDE experiencers and ghost-hunters remains an insistence that afterlife *is* provable, itself a kind of psi phenomenon, that evidence is abundant to those with eyes to see and ears to hear. The burden of proof, however, seems to fall back as ever upon subjective criteria of evidence and the reliability of fallible testimony.

Ψ

This realm has been given vivid visual expression in such semi-sf movies as *Flatliners* (1990), where young medical students deliberately induce NDEs and apparently suffer a blurring of their worlds between this and the next. In Natalie Wood's final movie, director and special effects genius Douglas Trumbull's *Brainstorm* (1983), a scanner captures and displays the transition between death and afterlife. *White Noise* (2005) draws upon the idea that hissing tape noise might be modulated by the dead to convey barely detectable messages to the living.

In *What Dreams May Come* (1998), based on Richard Matheson's novel, a man follows his dead and psychically wounded wife through a harrowing of hell in a quite astonishing blend of bathos and pathos. For Kevin Costner in the silly but oddly haunting *Field of Dreams* (1989), ghosts of eight disgraced Chicago White Sox ball players are summoned when he builds a baseball field in his Midwest farm. *Ghost* (1990) is street-parapsychological in tenor. *Chances Are* (1989) has a dead husband return as the boyfriend of his wife's daughter, and then fall in love with the still grieving widow; in 2004, Nicole Kidman in *Birth* carried this to an even creepier level, persuaded that a 10-year-old boy is her dead husband returned.

Are these psience fiction? Probably not quite; there is usually no enabling rationalization in terms of psi powers. Still, such movies perhaps might not have their impact without the prior immense success of fantastika tropes in the cinema, and the sf-based special effects that enliven fairytales into the dazzling representational realism of cinema.

Ψ

Back here in our shared empirical world, as Sargent and Watson noted astutely decades ago, "Nationalist politics as mediated by Islamic clergy has now given us modern religious martyrs, while in the most advanced technological country many millions of fundamentalist Christians view nuclear Armageddon with positive enthusiasm, given that those who believe will be 'raptured' to heaven" (1986, xiii). Since that observation, and especially after 9/11 and subsequent numerous mass murders in the name of Allah, terrorist "martyrs," convinced that they go after death to a banal paradise of 72 virgin females for every man, have wrought horror in that technological heartland, and "martyr brigades" strike with their own exploded flesh, or trucks surging into crowds, against foes in Israel, Iraq, France, Britain, and other nations where opposed conglomerates of money powered by faith, and vice versa, battle furiously for supremacy. This kind of dire slow-burn Armageddon is driven by people anguished by mean lives on earth and hoping desperately for a better existence beyond death. It has rarely been treated with any nuance in sf or psience fiction.

The general liberal or Enlightenment background assumption, after all, has been that a belief in afterlife will wither away as technology serves up utopia. While it is true—despite rumors to the contrary—that more people now live more secure and comfortable lives than ever before in history, with lifespans generally increasing in the privileged parts of the world, a suspicion grows that wealthy Westerners thrive only at the expense of the rest, including the poor or workless in their own nations, and at the cost of a world rushing into Greenhouse and resource-depletion horror. Reincarnation might prove less tempting as a belief as that suspicion hardens. Steely disbelief, though, is a stoic virtue perhaps beyond the reach of suffering people who defer their hopes to a better life beyond the grave.

Unless mediums and parapsychologists demonstrate unequivocally that such a domain is real and attainable, its adherents will regard afterlife as something to be hoped for in private faith, rather than by water-tight public evidence. That is a posture somewhat antithetical to the spirit of science fiction. As more desperados rich and poor blow themselves and their victims apart, we might expect sf to turn away in resolute revulsion from the premise of spiritual afterlife, and emphasize instead the kinds of technofixes cited above: cryonic suspension in expectation of eventual medical repair, rejuvenation, indefinite life extension, uploads, a transformative transition to some posthuman condition we can scarcely yet imagine. If science and psience fiction is right, Death will be defeated rather than embraced or railed against uselessly, so that after life we can expect more life, in this world however transcended, and after that more, and still more, perhaps to the re-ignition or budding off elsewhere of a drained and failing cosmos.

Chapter Notes

Preface

1. The preferred abbreviation of "science fiction" among professional writers is still "sf"—as in the title of *The Magazine of Fantasy & Science Fiction*, known for generations as *F&SF*, and in the title of many annual anthologies of "The Year's Best SF."
2. L. Flood, "Book Reviews," *New Worlds* 50, August 1956, pp. 126–28.
3. They were *not* taught to stare at goats until the luckless beasts died (as a 2004 movie starring George Clooney pretended, as did the flip Jon Ronson book it was based on).

Chapter 1

1. *Blue Book Magazine*, January 1955; *The World Treasury of Science Fiction*, ed. David G. Hartwell (Little, Brown, 1989), p. 613.
2. P. Bancel, "An Analysis of the Global Consciousness Project," in *Evidence for Psi*, eds. Broderick and Goertzel (McFarland, 2015), pp. 255–277.
3. http://www.queensu.ca/encyclopedia/h/humphrey-george.

Chapter 2

1. "Far Centaurus," *Astounding*, January 1944, reprinted in *Destination Universe!* by A.E. van Vogt (Panther Books, 1960), p. 20.

Chapter 3

1. John Clute, "Foreword" to the 1998 Old Earth Books editions of Smith's revisions.

2. John Clute, http://www.sf-encyclopedia.com/entry/smith_e_e.

Chapter 4

1. Paul Meehan, in *Cinema of the Psychic Realm*, mistakenly reports that the psychic slans have "a head endowed with a mass of tendrils that enable telepathy in place of hair" (p. 22). This Medusa effect is not supported by the text.
2. http://www.icshi.net/sevagram/summaries/slan1-preface.php#top.
3. This Nazi term remains in later revisions.
4. The book is suitably, if repetitively, excoriated by Isaac Walwyn in "The Young and the Tendrilless: A Golden Age Soap Opera," http://www.icshi.net/sevagram/reviews/slan-hunter2.php#top.
5. In the year Korzibski died, van Vogt embraced the doctrine of Dianetics, pulled together and propounded with enormous confidence and charisma by his fellow *Astounding* writer L. Ron Hubbard—going so far as to set up a practice in this dogma fated to become a particularly ludicrous religion, although by then van Vogt, eccentric but no fool, had bailed out.

Chapter 5

1. Sam Moskowitz, "The First College-Level Course in Science Fiction," *Science Fiction Studies*, 70 (Volume 23, Part 3), November 1996; http://www.depauw.edu/sfs/backissues/70/moskowitz70art.htm.
2. R.A.W. Lowndes, "Introduction," *Jack of Eagles* (Avon, 1982), p. 7.

3. Brian M. Stableford, *A Clash of Symbols: The Triumph of James Blish* (Borgo, 1979).
4. Damon Knight, *The Futurians* (John Day, 1977), p. 152.

Chapter 7

1. http://babylon5.wikia.com/wiki/Alfred_Bester.

Chapter 8

1. A story submitted shortly before Henderson's death more than a third of a century ago to Harlan Ellison's fabled and still unpublished anthology *Last Dangerous Visions*. It is unclear whether Ellison accepted the story, but it remained in limbo until NESFA published *Ingathering* (information provided by Mark Olson, May 11, 2017).
2. http://eom.byu.edu/index.php/God_the_Father.
3. R. Allbery, https://www.eyrie.org/~eagle/reviews/books/0-380-01507-2.html and https://www.eyrie.org/~eagle/reviews/books/0-380-01506-4.html.

Chapter 9

1. See *Building New Worlds*, by John Boston and Damien Broderick (Borgo, 2013), pp. 88–90. This information is cited from a personal communication from Ian Covell who obtained it from McIntosh in the course of preparing a bibliography of his work.

Chapter 11

1. https://www.theguardian.com/books/booksblog/2008/jan/29/; aliteraryargumentagainstde, Jan. 29, 2008.
2. "Afterword," *The Science Fiction of Mark Clifton* (Southern Illinois University Press, 1980), p. 294.
3. *Astounding* shared the first with *Galaxy*, and won the second outright.

Chapter 13

1. Tucker's pranksterish ways pepper *Wild Talent*'s cast with the names of other sf writers and editors such as Carnell, magazine publisher of this very novel. That rather irritating gag has been given the name "Tuckerization" by sf fans. Cf. https://fanlore.org/wiki/Tuckerized.
2. In *The Ultra Secret*, by Frederick Winterbottham.

Chapter 14

1. It is possible that Schmitz is making a sly reference here to L. Ron Hubbard's Dianetics, then recently rebranded as a religion, Scientology, where such mental exercises are performed. It is also possible that this novella appeared in *Galaxy* rather than *Astounding*, where much of his fiction appeared before and after this time, because Campbell was offended by the parallel.

Chapter 15

1. See the analysis in *Science Fiction: The 101 Best Novels, 1985–2010*, by Damien Broderick and Paul Di Filippo.
2. *The Chrysalids*, [1955] 2008, *New York Review of Books*, p. 5.

Chapter 18

1. Cf. the Tristero System in Thomas Pynchon's early triumph, from 1966, *The Crying of Lot 49*.

Chapter 19

1. Detailing the complex and endlessly entertaining plot of the novel is irrelevant for our purposes. For anyone who has never read *The Stars My Destination*, my advice is to do so at once. Short of that, see the badly written and necessarily oversimplified synopsis at http://www.sciencefictionmuseum.com/stories/reviews/snop009.html.
2. In the handsome but error-clogged British hardcover edition of *Tiger! Tiger!* from Wendover: John Goodchild Publishers, this paragraph is mysteriously redacted. Annoyingly, the male and female astrological characters in names such as Moira and Joseph and Nomad (a tiny cross beneath the

O in female names, an upward arrow above male names) are also absent.

Chapter 21

1. For a detailed list of these novels and anthologies, and links to summaries, see https://en.wikipedia.org/wiki/List_of_Darkover_books.
2. See, e.g., https://www.scientificamerican.com/article/why-does-the-brain-needs/.
3. See Jeffrey Mishlove, PhD, *The PK Man: A True Story of Mind Over Matter* (Hampton Roads, 2000).
4. http://darkover.wikia.com/wiki/Main_Page. That site has been of great use to me in preparing this chapter.

Chapter 22

1. Citations to "Parapsyche" are drawn from the ebook version of the collection, which has no page numbers, but quotations can be located in the ebook by using the search function.

Chapter 23

1. kornbluthcm-themindworm-00-t.txt
2. http://www.gutenberg.org/ebooks/34420.

Chapter 25

1. To use a derisive British term for a man who is old-fashioned cis-masculine to a sexist fault.
2. See, e.g., Russell Targ (2012), Joseph McMoneagle (2002), Paul Smith (2005).

Chapter 27

1. The full novella was out of print for decades before it was rescued in 2015 as the title story in a collection of short fiction from John Carnell's *Science Fantasy* magazine, edited by Damien Broderick and John Boston.

Chapter 29

1. It can be viewed at http://lowres-picturecabinet.com.s3-eu-west-1.amazonaws.com/53/main/23/174108.jpg.

Chapter 31

1. Dick, *The Father-Thing: The Collected Stories*, vol. 3 (Gollancz, 1989), pp. 373–74, cited from a 1979 address. Ironically, in 1978 his novel *Flow My Tears, the Policeman Said* was awarded the John W. Campbell Memorial Award, a very Dickian outcome of history.
2. Note on the *Galaxy* story "Oh, to Be a Blobel" (1964) reprinted in *The Days of Perky Pat: The Collected Stories*, vol. 4 (Gollancz, 1990), p. 379.
3. Cf. http://skeptiko.com/257-diane-powell-telepathy-among-autistic-savant-children/.

Chapter 32

1. See this illustrated in Russell Targ, *The Reality of ESP* (2012), pp. 62, 110–02, 119.

Chapter 33

1. I am not making this up as a parody; see https://www.glaad.org/reference/offensive: "Offensive: 'homosexual' (n. or adj.); Preferred: 'gay' (adj.); 'gay man' or 'lesbian' (n.); 'gay person/people'; Please use gay or lesbian to describe people attracted to members of the same sex. Because of the clinical history of the word 'homosexual,' it is aggressively used by anti-gay extremists to suggest that gay people are somehow diseased or psychologically/emotionally disordered– notions discredited by the American Psychological Association and the American Psychiatric Association in the 1970s. Please avoid using 'homosexual' except in direct quotes. Please also avoid using 'homosexual' as a style variation simply to avoid repeated use of the word 'gay.' The Associated Press, *The New York Times* and *The Washington Post* restrict use of the term 'homosexual.'"
2. Anyone interested in seeing Delany's reading of the novel at full throttle should go to "The Order of 'Chaos,'" in *Science Fiction Studies* 19 (Vol. 6, Part 3), November

1979: https://www.depauw.edu/site/sfs/reviews_pages/r19.htm#A19.
3. Theodore R. Cogswell, "Limiting Factor," *Galaxy*, April 1954, pp. 59–70.

Chapter 35

1. Michael Dirda, *The Washington Post*, http://www.washingtonpost.com/wp-dyn/content/article/2009/04/08/AR2009040803663.html
2. Personal communication, cited in Broderick, *Outside the Gates of Science*, p. 237.

Chapter 36

1. http://jamesdavisnicoll.com/review/who-watches-the-watchmen.

Chapter 37

1. Cornell Emeritus Professor Daryl J. Bem showed exactly this in several remarkable time-reversed experiments, where reinforcement of the correct choice was delivered *after* rather than *prior to* a test. Despite rote claims by skeptics that these "Feeling the Future" experiments have never been successfully replicated, they have been, and significantly often. See Bem 2011, 2015.

Chapter 38

1. T.M. Wagner: http://www.sfreviews.net/mindofmymind.html.

Chapter 41

1. Mentis is a real surname, as it happens, but it seems rather cheesy to give it to a telepath.
2. Max Dashú, "Woman Shaman," http://www.suppressedhistories.net/articles/womanshaman.html.

Chapter 42

1. http://www.hprg.com/athey/files/haplogroups.htm.
2. D. Broderick, *Reading by Starlight:*

Postmodern Science Fiction (Routledge, 1995), p. 61.
3. Tony Jinks, *An Introduction to the Psychology of Paranormal Belief and Experience* (McFarland, 2012), p. 34.

Chapter 43

1. See discussion of their colleague and friend Ingo Swann with Dr. Jacques Valleé, Dr. Harold Puthoff, Lt. Col. (ret.) Thomas McNear, Swann archivist Blynne Olivieri, and filmmaker Maryanne Bilham, moderated by Daniel Abella: https://www.youtube.com/watch?v=-qAqwA_5UPE.
2. http://www.ingoswann.com/human-genome.html.
3. Textbookx.com, 1980.
4. http://fmwriters.com/Visionback/Issue30/marysue.htm.
5. See Swann's site, BioMindSuperPowers.com, *Superpowers of the Human Biomind*.
6. This claim has been made many times, frequently by the stage conjurer the Amazing Randi (until lawsuits by Geller cost him a lot of money), but consider this datum: "[Geller] hosted a reality show called *The Successor*, which was a talent search for his psychic heir. On one episode he was caught cheating with a magnet on his fingertip to move a compass needle. This became a huge scandal, especially in Israel, where he was pummeled by the press. Geller claimed that he had good days and bad days, and someone suggested that he felt pressure to cheat on his bad days" (Diane Hennacy Powell, M.D., *The ESP Enigma: The Scientific Case for Psychic Phenomena* [Walker, 2009], p. 96).
7. Cf. William G. Roll, *The Poltergeist* (London: Wyndham, 1972/1976); see also Michaeleen Maher, "Ghosts and Poltergeists," Chap. 25, *Parapsychology: A Handbook for the 21st Century*, ed. Etzel Cardeña, et al. (McFarland, 2015), pp. 327–340. See also Guy Lyon Playfair, *This House is Haunted* (1980) about the so-called "Enfield poltergeist." Playfair (1935—) is the co-author, with Geller, of *The Geller Effect* (1986).
8. Writer and psi researcher Playfair, a school friend of the late sf master John Brunner, assures me that Geller wrote this book alone.

Chapter 44

1. One such is cited in Jonathan Margolis, *Uri Geller: CIA Masterspy* (2013), p. 108.

Conclusion

1. See an argument to this effect in https://www.theatlantic.com/science/archive/2016/09/panpsychism-is-wrong/500774/.
2. See the very funny but nicely researched chapter on this crazy idea in Mary Roach's *Spook: Science Tackles the Afterlife* (Norton, 2005), Chapter 2, "The Little Man Inside the Sperm...," pp. 57–75.
3. See David Luke, "Drugs and Psi Phenomena," in *Parapsychology: A Handbook for the 21st Century*, eds. Etzel Cardeña, et al. (McFarland, 2015), pp. 149–164.
4. Roach, p. 295.

Appendix 2

1. See, for example, Gerald A. Larue, professor emeritus of religion and adjunct professor of gerontology, University of Southern California, "Afterlife," *Humanism Today*, vol. 5 (1989).
2. See http://www.ibiblio.org/jstrout/uploading/ and the expert explorations in *Intelligence Unbound: The Future of Uploaded and Machine Minds*, ed. Damien Broderick and Russell Blackford (Wiley Blackwell, 2014).
3. See my *The Spike*, 2001, and various papers in *Intelligence Unbound*, for a longer discussion.
4. http://www.zenit.org/english/visualizza.phtml?sid=62884.
5. See my diptych *God Players*, 2005, and *K Machines*, 2006.
6. http://www.simulation-argument.com/.
7. Bostrom, *Superintelligence: Paths, Dangers, Strategies* (Oxford University Press, 2014).
8. For example, a high-tech but barbarian culture in my novel *The White Abacus* (1997) maintains that the soul is anchored to the body via the vermiforme appendix, which helps explain the more deplorable excesses of the 20th and 21st centuries when appendectomies became commonplace. In my *Omni* story "Thy Sting" (1987), study of the full genome reveals that many introns, apparently "junk" genetic intrusions, are actually a repository of living ancestral experiences, perhaps non-locally connected to their originals, permitting partial recovery of memories from our forebears.
9. With Silverberg's permission, I published a long continuation of that story, "Quicken," which was joined with his as the novel *Beyond the Doors of Death* (2013).
10. That is certainly the background of a key character in my novel *Transcension* (2002), who is sliced and diced and reconstructed, eventually forming part of the core description of the Aleph, a machine intelligence of preternatural power. It is the Aleph's decision to store all the world's people, if they agree (as most do), in simulated worlds prior to escaping the limitations of our four dimensional cosmos for somewhere roomier and more hospitable. This is an afterlife to rival those of the old pagans, but it lacks the truly transcendental (and paradoxical) character of contemporary faiths.

References

Entries are by author, in chronological order by first publication.

Aldiss, Brian W. "Psyclops." *New Worlds* 49, July 1956.
Anderson, Poul. "Journeys End." *Magazine of Fantasy & Science Fiction*, February 1957.
_____. "The Martyr." *Magazine of Fantasy & Science Fiction*, March 1960.
_____. "Night Piece." *Magazine of Fantasy & Science Fiction*, July 1961.
Asimov, Isaac. "Belief." *Astounding*, 1953. Original text in Grafton, *The Alternate Asimovs* (1987).
Bancel, Peter. "An Analysis of the Global Consciousness Project." In *Evidence for Psi*, eds. Damien Broderick and Ben Goertzel (McFarland, 2015).
Bear, Greg. *Queen of Angels*. Gollancz, 1990.
Bem, Daryl J., et al. "Feeling the Future: Experimental Evidence for Anomalous Retroactive Influence on Cognition and Affect." *Journal of Personality and Social Psychology* (March 2011) 1003: 407–25. doi: 10.1037/a0021524 ; *F1000Research* 2015, 4:1188 doi: 10.12688/f1000research.7177.1 .
Beresford, J. D. *The Hampdenshire Wonder*. Sidgwick & Jackson, 1911.
Berger, Albert I. *Science-Fiction Studies*, vol. 6, part 2 (1979), http://www.depauw.edu/sfs/backissues/18/berger18art.htm.
Bester, Alfred. *The Demolished Man*. *Galaxy*, January-March, 1952; Penguin, 1966.
_____. *The Stars My Destination*. *Galaxy*, 1956–57; Berkley Medallion, 1976; as *Tiger! Tiger!*, Penguin, 1967.
Blish, James. *Jack of Eagles*. Greenberg, 1952; Arrow, 1975. Expansion of "Let the Finder Beware," *Thrilling Wonder Stories*, December 1949.
Boston, John, and Damien Broderick. *Building New Worlds*. Borgo, 2013.
Bostrom, Nick. *Superintelligence: Paths, Dangers, Strategies*. Oxford University Press, 2014.
Bradley, Marion Zimmer. *The Planet Savers* [1958]. Ace Double, 1962.
Broderick, Damien. *The Dreaming Dragons*. Pocket, 1980. Revised and updated as *The Dreaming*, Fantastic, 2009.
_____. *Reading by Starlight: Postmodern Science Fiction*. Routledge, 1995.
_____. *Outside the Gates of Science*. Thunder's Mouth, 2007.
_____, and Paul Di Filippo. *Science Fiction: The 101 Best Novels, 1985–2010*. Nonstop, 2012.
Broderick, Mick. *Nuclear Movies*. McFarland, 1991.
Brunner, John. *Times Without Number*. Ace, 1969.
_____. *Telepathist/The Whole Man*. [Ballantine, 1964] Fontana, 1978.
Budrys, Algis. "Riya's Foundling." *Science Fiction Stories* 1, 1953.
Butler, Octavia E. *Patternmaster*. Doubleday, 1976.
_____. *Mind of My Mind*. Doubleday, 1977.
_____. *Survivor*. Doubleday, 1978.
_____. *Wild Seed*. Doubleday, 1980.

———. *Clay's Ark*. St. Martin's, 1984.
———. *Parable of the Sower*. Four Walls Eight Windows, 1993.
———. *Parable of the Talents*. Seven Stories, 1998.
Campbell, John. "The Science of Psionics." Editorial, *Astounding*, February 1956.
———. "The Problem of Psionics." Editorial, *Astounding*, June 1956.
———. "We *Must* Study Psi." Editorial, *Astounding*, January 1959.
———. "The Space-Drive Problem." *Astounding/Analog*, June 1960.
———. *The John W. Campbell Letters*, vol. 1. Eds. Perry A. Chapdelaine, et al. AC Projects, 1985.
Cardeña, Etzel, et al., eds. *Parapsychology: A Handbook for the 21st Century*. McFarland, 2015.
Cherryh, C. J. "Cassandra." *Magazine of Fantasy & Science Fiction*, October 1978.
Clarke, Arthur C. *Childhood's End*. Ballantine, 1953.
———. *The City and the Stars*. Harcourt, Brace, 1956.
———. *2001: A Space Odyssey*. New York: Signet, 1968.
Clifton, Mark. "Star, Bright." *Astounding*, July 1952.
———. "Sense from Thought Divide." *Astounding*, March 1955.
———. "How Allied." *Astounding*, February 1957.
———. "Remembrance and Reflection." *Magazine of Fantasy & Science Fiction*, January 1958.
———. *Pawn of the Black Fleet/When They Come From Space*. Amazing, 1961–62.
Clifton, Mark, and Frank Riley. *They'd Rather Be Right*. *Astounding*, August, September, October, November 1954; Starblaze edition (Donning, 1981).
Clifton, Mark, with Alex Apostolides. "What Thin Partitions." *Astounding*, September 1953.
Clute, John, and Peter Nicholls. *Encyclopedia of Science Fiction*, 2d ed., Orbit, 1993.
Clute, John, et al. *SFE: Encyclopedia of Science Fiction*, 3d ed., http://www.sf-encyclopedia.com/.
Cogswell, Theodore R. "Limiting Factor." *Galaxy*, April 1954, pp. 59–70.
Condon, Richard. *The Manchurian Candidate*. McGraw-Hill, 1959.
Cowper, Richard John Middleton Murry, Jr. *Breakthrough*. Dennis Dobson, 1967.
Dashú, Max. "Woman Shaman." http://www.suppressedhistories.net/articles/womanshaman.html.
Delany, Samuel R. *The Jewel-Hinged Jaw: Notes on the Language of Science Fiction*. Dragon, 1977.
———. "The Order of 'Chaos.'" In *Science Fiction Studies* 19, vol. 6, part 3, November 1979.
Del Rey, Lester. "For I Am a Jealous People." In *Star Short Novels*, ed. Frederik Pohl (Ballantine, 1954).
———. *Pstalemate*. Berkley, 1971.
Dick, Philip K. "Imposter." *Astounding*, June 1953.
———. "A World of Talent." *Galaxy*, October 1954.
———. "Psi Man Heal My Child!" *Imaginative Tales*, November 1955.
———. "The Minority Report." *Fantastic Universe*, January 1956.
———. *Eye in the Sky*. Ace, 1957.
———. *Do Androids Dream of Electric Sheep?* Doubleday, 1968.
———. *Ubik*. Doubleday, 1969.
———. *The Collected Stories of Philip K. Dick*, vol. 3. Gollancz, 1989.
———. *The Collected Stories of Philip K. Dick*, vol. 4. Gollancz, 1990.
Dunne, J. W. *An Experiment with Time*. Faber & Faber, 1927.
Egan, Greg. "Learning to Be Me." *Interzone*, July 1990.
———. *Permutation City* Millennium. 1994.
———. *Diaspora* Millennium. 1997.
Farmer, Philip José. *To Your Scattered Bodies Go*. Putnam, 1971.
———. *The Fabulous Riverboat*. Putnam, 1971.
———. *The Dark Design*. Berkley/Putnam, 1977.
———. *The Magic Labyrinth*. Berkley/Putnam, 1980.
———. *Gods of the Riverworld*. Putnam, 1983.
Flood, Leslie. "Book Reviews." *New Worlds* 50, August 1956.

Galouye, Daniel F. *Simulacron-3*. Bantam, 1964.
Geller, Uri. *Ella*. Headline, 1998.
Godwin, Tom. "The Cold Equations." *Astounding*, August 1954.
Gold, H. L. Editorial. *Galaxy*, January 1956.
Hanson, Robin. *The Age of Em: Work, Love and Life When Robots Rule the Earth*. Oxford University Press, 2016.
Heinlein, Robert A. *Beyond This Horizon* [1942, 1948]. New English Library, 1978.
_____. *The Puppet Masters*. Pan, 1969.
_____. *Time for the Stars* [1956]. Pan, 1978.
_____. *I Will Fear No Evil*. Putnam, 1970.
_____. *The Number of the Beast*. New English Library, 1980.
Henderson, Zenna. *Ingathering: The Complete People Stories*. NESFA, 1995.
Hamilton, Peter F. *The Reality Dysfunction*. Pan, 1996.
_____. *The Neutronium Alchemist*. Pan, 1998.
_____. *The Naked God*. Pan, 2000.
Hubbard, L. Ron. "To the Stars." *Astounding*, February-March, 1950.
James, Edward. "Utopias and Anti-Utopias." In *The Cambridge Companion to Science Fiction*, 2003.
Jinks, Tony. *An Introduction to the Psychology of Paranormal Belief and Experience*. McFarland, 2012.
Koch, Christopher J. Review, *Sydney Morning Herald*, 3 October 1979.
Knight, Damon. *In Search of Wonder*. Advent, 1967.
_____. "Knight Piece." In *Hell's Cartographers*, eds. Brian W. Aldiss and Harry Harrison [1975] (Orbit, 1976).
_____. *The Futurians*. John Day, 1977.
_____. "The Tree of Time"/*Beyond the Barrier*. Magazine of Fantasy & Science Fiction, 1963; Macfadden-Bartell, 1965.
Kornbluth, C.M. "The Mindworm." *Worlds Beyond*, December 1950.
Kuttner, Henry, and Catherine L. Moore as Lawrence O'Donnell. "Fury." *Astounding*, May-July, 1947.
_____ as Lewis Padgett. *Mutant* series. *Astounding*, 1945–53.
Le Guin, Ursula K. *The Farthest Shore*. Gollancz, 1973.
Lem, Stanislaw. "Philip K. Dick: A Visionary Among the Charlatans" [1975]. In *Microworlds: Writings on Science Fiction and Fantasy*, ed. Franz Rottensteiner (Secker & Warburg, 1985), pp. 106–135.
Lowentrout, Peter M. "PsiFi: The Domestication of Psi in Science Fiction." *Extrapolation* 304 (Winter 1989): 388–403 and at http://web.csulb.edu/~plowentr/Psifi.htm.
Malzberg, Barry N. *The Engines of the Night: Science Fiction in the Eighties*. Doubleday, 1982.
_____, ed. *The Science Fiction of Mark Clifton*. Southern Illinois University Press, 1980.
May, Edwin C., and Sonali Bhatt Marwaha, eds. *Anomalous Cognition: Remote Viewing Research and Theory*. McFarland, 2014.
_____. *Extrasensory Perception: Support, Skepticism, and Science*, two vols. Praeger, 2015.
_____. *The Star Gate Archives: Reports of the United States Government Sponsored Psi Program, 1972–1995*, four vols. McFarland, 2018 and forthcoming.
MacLean, Katherine. *Missing Man*. Berkley, 1975.
_____. "Defense Mechanism." *Astounding*, October 1949.
_____. "Incommunicado." *Astounding*, June 1950.
_____. "Pictures Don't Lie." *Galaxy*, August 1951.
_____. "The Snowball Effect." *Galaxy*, September 1952.
MacLeod, Ken. *Newton's Wake*. Tor, 2004.
Macpherson, Donald. *Go Home, Unicorn*. Faber & Faber, 1935.
McCaffrey, Anne. "A Womanly Talent." *Analog*, February 1969.
_____. *To Ride Pegasus*. Ballantine, 1973.
_____. *Pegasus in Flight*. Del Rey/Ballantine, 1990.

_____. *Pegasus in Space*. Del Rey/Ballantine, 2000.
McCarthy, Cormac. *The Road*. Alfred A. Knopf, 2006.
McFadden, Johnjoe. *Quantum Evolution: The New Science of the Life Force*. HarperCollins, 2000.
McIntosh, J. T. *The ESP Worlds*. New Worlds 16–18, July-November 1952.
_____. "Empath." *New Worlds* 50, August 1956.
McMoneagle, Joseph. *The Stargate Chronicles: Memoirs of a Psychic Spy*. Hampton Roads, 2002.
Meehan, Paul. *Cinema of the Psychic Realm: A Critical Survey*. McFarland, 2009.
Merril, Judith. "Theodore Sturgeon." *Magazine of Fantasy & Science Fiction*, September 1962.
Miller, Walter M., Jr. "Command Performance." *Galaxy*, November 1952.
Mishlove, Jeffrey. *The PK Man: A True Story of Mind Over Matter*. Hampton Roads, 2000.
Morgan, Dan. *The New Minds*. Corgi, 1967.
_____. *The Several Minds*. Avon 1969.
_____. *Mind Trap*. Avon, 1970.
_____. *The Country of the Mind*. Corgi, 1975.
Nevala-Lee, Alec. *Astounding: John W. Campbell, Isaac Asimov, Robert A. Heinlein, L. Ron Hubbard, and the Golden Age of Science Fiction*. Dey Street, forthcoming.
Panshin, Alexei, and Cory Panshin. *The World Beyond the Hill: Science Fiction and the Quest for Transcendence*. Tarcher, 1989.
Phillips, Mark, Randall Garrett and Laurence M. Janifer. *Brain Twister* [1962]/"That Sweet Little Old Lady,"*Astounding/Analog*, September-October 1959.
_____. *The Impossibles*. Pyramid [1963]/"Out Like a Light," *Astounding/Analog*, January, May, June 1960.
_____. *Supermind*. Pyramid [1963]/"Occasion for Disaster," *Analog*, November 1960, January, February 1961.
Pilkington, Rosemarie, ed. *Men and Women of Parapsychology: Personal Reflections*. McFarland, 1987.
Piper, H. Beam. "Last Enemy." *Astounding*, August 1950.
Playfair, Guy Lyon. *This House Is Haunted: The True Story of the Enfield Poltergeist*. Stein and Day, 1980.
Pohl, Frederik. *The Other End of Time*. Tor, 1996.
_____. *The Siege of Eternity*. Tor, 1997.
_____. *The Far Shore of Time*. Tor, 2000.
Powell, Diane Hennacy, M.D. *The ESP Enigma: The Scientific Case for Psychic Phenomena*. Walker, 2009.
Pullman, Philip. *The Amber Spyglass*. Knopf, 2000.
Pynchon, Thomas. *The Crying of Lot 49* [1966]. Bantam, 1967.
Robinson, Frank M. *The Power* [1956]. Revised, Tor, 2000.
_____. *Waiting*. Forge, 1999.
Robinson, Spider. *Time Pressure*. Ace, 1987.
_____. *Lifehouse*. Baen, 1997.
Roll, William G. *The Poltergeist*. Star/Wyndham, 1976.
Rush, Joseph H. "Parapsychology: Some Personal Observations." In Pilkington, 1987, pp. 60–73.
Russ, Joanna. *And Chaos Died*. Ace Special, 1970.
_____. *The Female Man*. Bantam, 1973.
Russell, Eric Frank. *Sinister Barrier Unknown* [1939]. Methuen, 1986.
Sargent, Pamela, and Ian Watson, eds. *Afterlives*. Vintage, 1986.
Schmitz, James H. *The Ties of Earth*. *Galaxy*, November 1955-January 1956.
_____. "The Witches of Karres." *Astounding*, December 1949.
_____. *The Witches of Karres* [1966], VGSF, 1988.
Sellings, Arthur. *Telepath*. Ballantine, 1962.
Shaw, Bob. *The Palace of Eternity*. Ace SF Special, 1969.

Sheckley, Robert. *Immortality, Inc.* Bantam, 1959.
Shepard, Lucius. *Life During Wartime* [1987], Paladin, 1989.
Silverberg, Robert. *Dying Inside* [1972], Ballantine, 1973.
____. "Something Wild Is Loose" [1971]. In *The Collected Stories of Robert Silverberg*, vol. 3, Subterranean, 2008.
____. "Born with the Dead." *Magazine of Fantasy & Science Fiction*, 1974.
____. *The Stochastic Man. Magazine of Fantasy & Science Fiction*, 1975.
Simak, Clifford. *Time and Again.* Simon & Schuster, 1951.
____. *Time Is the Simplest Thing* as "The Fisherman," *Analog,* April-July 1961; Pan, 1962.
Smith, Cordwainer. "The Game of Rat and Dragon." *Galaxy,* October 1955.
____. "Alpha Ralpha Boulevard." *Magazine of Fantasy & Science Fiction,* June 1961.
Smith, E. E. *First Lensman* [1950]. Fantasy; Berkley, 1986.
____. *Galactic Patrol. Astounding,* September 1937-February 1938. Revised, Fantasy, 1950.
____. *Gray Lensman. Astounding,* October 1939-January 1940. Revised, Fantasy, 1951.
____. *Second-Stage Lensman. Astounding,* November 1941-February 1942. Revised, Fantasy, 1953.
____. *Children of the Lens. Astounding,* November 1947-February 1948. Revised, Fantasy, 1954.
Smith, George O. *Highways in Hiding.* Lancer, 1956.
Smith, Paul. *Reading the Enemy's Mind: Inside Star Gate—America's Psychic Espionage Program.* Forge, 2005.
Stableford, Brian. *A Clash of Symbols: The Triumph of James Blish.* Borgo, 1979.
____. "The Oedipus Effect." In *Temps,* eds. Neil Gaiman and Alex Stewart, 1981.
____. "Reincarnation." In *The Encyclopedia of Science Fiction,* eds. Peter Nichols and John Clute, pp. 999–1000 (Orbit, 1993).
Stapledon, Olaf. *First and Last Men* [1930]. Penguin, 1966.
____. *Odd John: A Story Between Jest and Earnest* [1935]. New English Library, 1978.
____. *Starmaker* [1937]. Penguin, 1972.
Stevenson, Ian, M.D. *Twenty Cases Suggestive of Reincarnation.* University Press of Virginia, 1980.
____. *Children Who Remember Previous Lives: A Question of Reincarnation,* rev. ed. McFarland, 2000.
Straczynski, J. Michael. *Babylon 5 Scripts of J. Michael Straczynski,* five volumes. Synthetic Worlds, November 2005-March 2006.
Stross, Charles. *Accelerando.* Ace, 2005.
Sturgeon, Theodore. "Baby Is Three." *Galaxy,* October 1952.
____. *More than Human* [1953]. Penguin, 1965.
____. "To Marry Medusa." *Galaxy* [1958]/*The Cosmic Rape,* Dell, 1958.
____. "When You Care, When You Love." *Magazine of Fantasy & Science Fiction,* 1962.
____. *Thunder and Roses,* vol. IV: *The Complete Stories of Theodore Sturgeon.* North Atlantic, 1997.
Swann, Ingo. *Star Fire.* Sphere, 1978.
Targ, Russell. *The Reality of ESP: A Physicist's Proof of Psychic Abilities.* Quest, 2012.
Tiedman, Richard. "Jack Vance: Science Fiction Stylist" in *Jack Vance,* eds. Tim Underwood and Chuck Miller (Taplinger, 1980).
Tipler, Frank. *The Physics of Immortality.* Anchor, 1995.
Tucker, Wilson. *Wild Talent. New Worlds* 26, August-October 1954.
Varley, John. "Overdrawn at the Memory Bank." *Galaxy,* May 1976.
van Vogt, A. E. *Slan Astounding.* September, October, November, December 1940; revisions 1946/Panther, 1968.
____. "Far Centaurus." *Astounding,* January 1944. Reprinted in *Destination Universe!,* Panther, 1960.
____. *The World of Null-A. Astounding,* August, September, October 1945; Simon & Schuster, 1948.

_____. *The Players of Null-A*. *Astounding*, October, November, December 1948, January 1949. Revised as *The Pawns of Null-A*, Ace, 1956.
_____. *Supermind*. DAW, 1977.
Vance, Jack. "Telek." *Astounding*, January 1952.
_____. "Parapsyche." *Amazing Stories*, August 1958.
_____. "The Miracle-Workers." *Astounding*, July 1958.
Vaughn, Carrie. *After the Golden Age*. Tor, 2011.
_____. *Dreams of the Golden Age*. Tor, 2013.
Vinge, Joan D. *Psion*. Delacorte, 1982.
_____. *Catspaw*. Warner, 1988.
_____. *Alien Blood* [*Psion* plus *Catspaw*]. SFBC, 1988.
_____. *Dreamfall*. Aspect/Warner, 1996.
Wagenmakers, Eric-Jan, et al. "A Skeptical Eye on Psi." In *Extrasensory Perception: Support, Skepticism, and* Science, vol. 1, Edwin C. May and Sonali Bhatt Marwaha, eds. (Praeger, 2015).
Walwyn, Isaac. "Slanology." http://www.icshi.net/sevagram/summaries/slan1-preface.php#top.
Wilson, Colin. *The Philosopher's Stone* [1974]. Granada, 1969.
_____. *The Mind Parasites* [Arkham House, 1967]. Bantam, 1968.
_____. *The Space Vampires*. Random House, 1976.
Willis, Connie. *Passage*. Bantam Spectra, 2001.
_____. *Crosstalk*. Del Rey, 2016.
Wiseman, Richard. Cited in http://www.dailymail.co.uk/news/article-510762/Could-proof-theory-ALL-psychic.html, 2008.
Woodcott, Keith John Brunner. *Crack of Doom*. *New Worlds* 122, September-October, 1962.
Wright, Lan. *A Man Called Destiny*. *New Worlds* 78, December 1958-February 1959.
Wyndham, John. *The Chrysalids*. Michael Joseph, 1955.
Zelazny, Roger. "He Who Shapes." *Amazing Stories*, 1964. Expanded as *The Dream Master*, Panther, 1968.

Index

absent signifieds 171
Accelerando 212
After the Golden Age 167, 187
afterlife 197, 207
Afterlives 214
The Age of Em 213
Aldiss, Brian W. 53, 105, 113, 187
Alien Blood 162
Allbery, Russ 50
Allen, Steve 24
"Alpha Ralpha Boulevard" 212
The Alternate Asimovs 101
Amazing Stories 21
The Amber Spyglass 214
Analog Science Fact-Science Fiction 16, 17, 108–9, 130, 152, 155, 188, 190
And Chaos Died 142–43, 187
Anderson, Kevin 36
Anderson, Poul 107, 178–81, 187, 211
The Angry ESPers 188
anomalous cognition 4, 197–98
anomalous perturbation 4, 132, 200
antipsi "inertials" 137
"Anybody Else Like Me?" 100
Apostolides, Alex 59, 64
"Ararat" 50
Asimov, Isaac 8, 70, 90, 101–3, 176, 185, 187
Asperger spectrum 154
Astounding Science Fiction 7–18, 28, 29, 32–33, 42, 59–60, 65–66, 74, 97–101, 108–9, 134, 152, 155, 166, 191, 209
"Asylum" 78
Atlas Shrugged 62

"Baby Is Three" 56
Babylon 5 46
Baldies 10
Ballantine Books 113, 145
Ballard, J.G. 53

Bancel, Peter 24
Barrett, Sir William F. 195
Bayesian statistics 202
Bear, Greg 16, 118
"Belief" 101, 176, 187
Benford, Gregory 208
Beresford, J.D. 26
Berger, Albert I. 9
Bergsonian life force 27
Bester, Alfred 4, 45–46, 49, 60–62, 69, 78, 83–87, 95, 98, 100, 111, 124, 154, 175, 184–88
Beyond the Occult 139
Beyond This Horizon 209
Bierman, Dick J. 201–2
Biggle, Lloyd 188
biological radio 18
Birth 10, 216
Black Easter/The Day After Judgment 180
Blackett-Dirac formula 40
Blackett equation 111
Bladerunner 215
bleshing 58
Blish, James 37–44, 109, 111, 136, 171, 180, 184–86
"Born with the Dead" 214
Bostrom, Nick 213
Boundary Institute 201
Bradley, Marion Zimmer 90–92, 161, 186
Brain Twister 108–9, 187
Brainstorm 216
Breakthrough 127, 187
Broughton, Richard 203
Budrys, Algis 103, 187
Butler, Octavia 158–61, 187

The Cambridge Companion to Science Fiction 25
Campbell, John W. 7–21, 28, 30, 33, 42,

229

Index

59–66, 97–101, 108–10, 134, 138, 152, 155, 166, 175, 185, 188, 190–91, 200–1
"capital-C" Consciousness 189
Captives of the Flame 115
Cardeña, Etzel 199–200
Carnell, John (Ted) 10, 53, 87, 115
Carpenter, James 203
Carr, Bernard 201
Carrie 167, 176, 184, 187
Carroll, Sean 2, 191
"Cassandra" 181, 187
Catspaw 162
Chances Are 216
Cherryh, C.J. 181, 187
Childe Cycle 188
Childhood's End 210
Children of Dune 28
Children of the Lens 28
Children Who Remember Previous Lives 209
Chrysalids 10, 72–73, 186
Chrysalids/Re-Birth 186
CIA 5, 11, 17, 49, 83, 199, 203–6
Cities in Flight 109
The City and the Stars 210
"City of the Tiger" 118
clairvoyance 2, 31, 39, 54, 80, 86, 99, 132, 142, 171
Clarke, Arthur C. 7, 59, 210
Clay's Ark 159
Clifton, Mark 59–66, 186, 191
Clute, John 29, 50, 208, 211
"cocktail party" effect 33
Coils 188
"The Cold Equations" 60
Coleman, Sidney R. 124, 126, 146
"Command Performance" 100, 181, 187
Condon, Richard 83
Cordwainer Smith 104
The Cosmic Rape 56
The Country of the Mind 121, 125, 163, 187
Cowper, Richard 127, 187
"Crazy Joey" 59
Crosstalk 169, 187–89
"curative telepathist" 118–20

The Dark Beyond the Stars 77
The Dark Design 211
Darkover series 90, 186
Davies, Paul 131
Dawkins, Richard 198
Dean Drive 17, 98
"Defense Mechanism" 98, 105, 152, 187
Delany, Samuel R. 4, 84, 115, 142, 143, 184
del Rey, Judy-Lynn 145
del Rey, Lester (Knapp, Leonard) 145, 187

The Demolished Man 4, 45–49, 78–84, 85, 91, 100, 109, 148, 184–86
DIA 5, 11
Dianetics 8, 12, 17, 62–63, 98, 152, 190–91
Diaspora 212
Dick, Philip K. 46, 82, 98, 103, 133–38, 183–88, 212–15
Dickson, Gordon 98, 188
Di Filippo, Paul 18
Dirda, Michael 149–50
Do Androids Dream of Electric Sheep 215
Dorsai! 188
double blind protocol 204
dowsers 4
Dragonriders of Pern series 90
Dreamfall 162
Dreams of the Golden Age 167
Duke University 13, 24–25
Dune 28, 36, 98, 188
Dunne, J.W. 41
Dying Inside 108, 149–51, 155, 187

Easter Bunny Research 2
Egan, Greg 212
egons 210
Eight Keys to Eden 64
ElectroGaiaGrams 130
Ella 61, 126, 137, 173–78, 190
Ellison, Harlan 164
emergees 66
"Empath" 106, 187
Emshwiller, Ed 55
Encyclopedia of Science Fiction 7
ESP Wars East and West 199
The ESP Worlds 52–53, 186
Esper 47- 49, 82, 152
esperance 62
espers 80
An Experiment with Time 41
extrasensory perception (ESP) 2–3, 7
Extrasensory Perception: Support, Skepticism, and Science (ed. Edwin May) 31, 199
Eye in the Sky 215

F&SF see *Magazine of Fantasy & Science Fiction*
The Fabulous Riverboat 211
The Fall of the Towers 115
Fans Are Slans 10, 61, 71
Fantastic Universe 134
"Far Centaurus" 27
The Far Shore of Time 211
Farmer, Philip José 211
Farnham's Freehold 161
The Farthest Shore 214

Index

Field of Dreams 216
First Sight theory of psi 203
Flatliners 216
Flood, Leslie 4, 85
"For I Am a Jealous People" 145
The Forever Machine 59
Fort, Charles 5, 11, 12
Fort Meade, Maryland (remote viewing) 5, 11
Frankenstein: The New Prometheus 21
"Fury" 10

Gaia hypothesis 70
Galactic Patrol 28–29
Galaxy Magazine 4, 8, 45–46, 56, 59–60, 69, 86–88, 100–1, 104, 134, 152, 215
Galouye, Daniel 188, 213
"The Game of Rat and Dragon" 104, 187
Garrett, Randall 108–9, 187
Gates, Robert 203, 204
Geller, Uri 61, 173–78, 190
General Semantics 37
genomics 73
Gernsback, Hugo 21
gestalt 11, 57–58, 65, 89, 144, 161–63
Ghost 200, 216
The Ghost in the Machine 200
The Glass Inferno 77
global consciousness field 66
Global Consciousness Project 24, 130
Go Home, Unicorn 21–25, 186, 189
The Godmakers 188
Gods of the Riverworld 211
Godwin, Tom 60
Gold, Horace L. 4, 60, 69, 134
Golden Age of Science Fiction 1, 8
Gould, Steven 131, 188
Gray Lensman 28, 42
guru 39, 124–25, 157

half-life 137, 213, 215
Hamilton, Peter F. 77, 209, 215
The Hampdenshire Wonder 26
Hanson, Robin 213
haplogroup 170
Harry Potter and the Sorcerer's Stone 138
"He Who Shapes" 118
Heinlein, Robert A. 7–8, 17, 73–77, 90, 161, 186, 209–10
Henderson, Zenna 11, 50–51, 67, 79, 186
Herbert, Frank 184, 188
Herodotus 205
"Hide! Hide! Witch!" 59
Hieronymus, T.G. 16, 98
Hieronymus machine 16
Highways in Hiding 80, 186

History of Civilization 28, 186
Homo gestalt 57, 58
Homo sapiens 10, 62, 69, 134, 192
Homo superior 25, 33, 69, 175, 179
"How Allied" 65
Hubbard, L. Ron 8, 12, 17, 74, 98, 191, 209
Humphrey, George William 21–24, 186
Huxley, Julian 22, 23
Hyman, Ray 198
hypnagogia 33
hypnopompia 33

I Will Fear No Evil 209
Imaginative Tales 134
Immortality, Inc 215
Impossibles 109, 187
"Imposter" 134
In Search of Wonder 100
Inception 213
Incommunicado 152
The Infinite Cage 188
The Infinite Mind 188
Introduction to the New Existentialism 139
Islands in the Sky 7, 184

"Jabberwocky" 191
Jack of Eagles 37, 109, 136, 171, 186
James, Edward 25
James, William 22, 123. 195
Janifer, Laurence M. 108, 187
John W. Campbell Letters, Volume 1 12
Jommy Cross 9, 34–36, 68
Jordison, Sam 60
Jothun, T.O. 15
Journal of Parapsychology 199, 202
"Journeys End" 107, 181, 187

Kelly, Edward F. 200
King, Stephen 176, 184
klatha magic 44
Knapp, Leonard see del Rey, Lester
Knight, Damon 8, 78, 83–84, 100
Koch, Christof 200
Koch, Christopher J. 140
Kornbluth, C.M. 99–100, 187
Korzybski, Count Alfred 37–38
Kuttner, Henry 10, 12, 190

laran 91, 92
Last and First Men 11, 21, 53
"The Last Enemy" 214
Laumer, Keith 71, 188
"Learning to Be Me" 212
Le Guin, Ursula 184, 214
Leiber, Fritz 142

Lem, Stanislaw 136, 138
Lensmen series 16, 178
"Let the Finder Beware" 37
levitation 13, 57, 101–2, 167, 177, 188
Life During Wartime 164–66, 187
Lifehouse 211
Linebarger, Paul Myron Anthony (Cordwainer Smith) 104
"literary" fiction 5
Lovecraft, H.P. 211
Lovelock, James 70
Lowentrout, Peter 5, 18, 191–92
Lowndes, Robert A.W. 38
Lurulu 98

MacLean, Katherine 98–99, 105, 152–55, 187
MacLeod, Ken 212
Macpherson, Donald 21, 25, 27, 186
Magazine of Fantasy & Science Fiction (aka *F&SF*) 8, 50, 55, 59, 66, 107, 155, 178–79, 181
magic 1, 4, 45, 158, 165–66, 186
The Magic Labyrinth 211
magical hand-waving 38
Malzberg, Barry 17, 60, 185
A Man Called Destiny 87, 186
The Manchurian Candidate 83
Many Worlds Theory 10
"The Martyr" 211
Marwaha, Sonali Bhatt 198–99, 201
Mass Cathexis 49, 86, 95
Matheson, Richard 216
The Matrix 213
May, Edwin C. 147, 173, 199–201, 206
McCaffrey, Anne 90, 98, 130–33, 159, 161, 187
McFadden, Johnjoe 131, 215
McIntosh, J.T. 52–55, 106–7, 186–87
McMoneagle, Joseph 31, 114, 140, 173, 201–6
mediums 4, 30, 94, 169, 195, 217
Mekstrom's disease 81–82
Men Are Like Animals 21, 25
Mengele, Josef 166
Merril, Judith 55, 66, 100, 106
metaphysical archetypes 191
metaphysical SF psi story 18, 192
Michaelmas 103
Miller, Walter M., Jr. 100, 181, 187
"Mimsy Were the Borogoves" 191
mind field 22
Mind of My Mind 158–61, 187
The Mind Parasites 138, 211
Mind Trap 121, 124, 187
"The Mindworm" 99–105, 187

"Minority Report" 46
"The Miracle Workers" 95, 186
Missing Man 152, 187
MK-Ultra experiments 49, 83
Moore, Catherine L. 10, 12, 190
More Than Human 52, 56–59, 65, 71, 186
Morgan, Dan 121–25, 130, 151, 163, 187, 215
morphic field 23
Moskowitz, Sam 37
Multiverse 38, 103, 157
Mutant 10
mutants 26, 32–33, 42, 68, 89, 134, 160
Myers, W.H. 195
Mysteries 139–40
myths 191

The Naked God 215
Near Death Experiences (NDEs) 189, 197
negative capability 127
Nelson, Roger 24
neuroscience 123
The Neutronium Alchemist 215
The New Minds 121, 187
New Worlds 4, 54, 87, 105–6, 115
Newton's Wake 212
Nicholl, James 155
"Night Piece" 178, 187
The Nightmare Factor 77
Noösphere 58
Nourse, Alan E. 188
Novik, Naomi 90
Nuclear Movies 9
Null-A Three 36
The Number of the Beast— 210

"Occasion for Disaster" 109, 111, 187
The Occult 139
Odd John 11, 19, 21, 24–27, 33, 42, 68, 186, 189
O'Donnell, Lawrence 10
"The Oedipus Effect" 182–83, 187
Omega Point 211
organ of Funck 119
The Other End of Time 211
"Out Like a Light" 109, 187
The Outsider 139
"Overdrawn at the Memory Bank" 212
Owens, Ted 92

Padgett, Lewis 10, 190
The Palace of Eternity 210
Palmer, John 199
panpsychism 189, 197
Panshin, Alexei and Cory 37, 189
Parable of the Sower 159
Parable of the Talents 159

parallel universes 38
paranormal transcendence 10
"Parapsyche" 92, 186
parapsychology 2, 196, 207
Parapsychology: A Handbook for the 21st Century 199
pareidolia 33, 200
Passage 189, 214
Patternmaster 159, 161
Pawn of the Black Fleet 63
The Pawns of Null-A 37
Pegasus in Flight 130, 133
Pegasus in Space 130, 133
The People 50, 186
Permutation City 212
Phillips, Mark 108-9, 187
The Philosopher's Stone 26, 129, 138, 140, 176, 187
The Physics of Immortality 211
Pictures Don't Lie 152
Pilgrimage 51
Piper, H. Beam 214
The Planet Savers 90
platting the twishers 51
The Players of Null-A 37
Playfair, Guy Lyon 176
Pohl, Frederik 99, 211-12
poltergeist 45, 64, 110, 141, 176, 177, 187
The Power 77-82, 90, 122, 186
precognition 3, 13, 24, 38-39, 43, 52, 67, 70-74, 86-87, 92, 128, 132, 135, 140, 145-48, 155-56, 165, 171, 181-83, 186, 201, 208
presentiment 193
Price, Pat 140
The Primal Urge 113
The Pritcher Mass 98, 188
"The Problem of Psionics" 15
The Prometheus Crisis 77
pseudoscience 2
psi (defined) 4
Psi High and Others 188
psi machines 16
"Psi Man Heal My Child!" 134-35
psi-mediation 53, 143
psi on the loose 111
psi research 186, 195
psience fiction 4 and *passim*
psinul 116-17
Psion 162-64, 187
psionic machine 38, 145, 213
The Psionic Menace 115, 116, 187
psionics 1, 4, 7, 12-17, 30, 62-64, 93, 134, 166, 188, 191
psions 100, 115-17, 144, 163
Pstalemate 145, 187
psychic field 22

psychic temporal loops 132
psychokinesis 3, 24, 49, 52, 91, 94, 131-33, 175-76, 186-87, 200
psychology of electrons 27
"Psyclops" 105, 187
"The Public Hating" 24, 49
Pullman, Philip 214
punctuated equilibrium 131
The Puppet Masters 77
Puthoff, Hal 3, 173, 206

quantum evolution 131
quantum theory 73, 195, 200, 203
Queen of Angels 118
Questionable Research Practices (QRPs) 202
Quick Frozen Foods 37

"R & R" 165-66
Radin, Dean 203
The Reality Dysfunction 215
"Remembrance and Reflection" 66
remote viewers 7, 67, 114, 121, 140, 173, 196, 204, 206
remote viewing 3, 17, 31, 86, 96, 99, 140, 165, 173, 192, 197, 203-6
"Rescue Squad" 152
Rhine, Joseph Banks 2-3, 12-15, 25, 57, 67, 79, 133, 150, 171, 175, 195-99
Rhine Institute 80
Riley, Frank 59-63, 186
"Riya's Foundling" 103, 187
Robinson, Frank M. 77-78, 89-90, 186
Robinson, Spider 211
Rogue Moon 103
Roll, William G. 176
Rosicrucianism 21
The Rowan 130, 133
Rush, Joseph A. 196-97
Russ, Joanna 142, 143-45, 184, 187
Russell, Eric Frank 3, 12-14, 43, 138, 206
Ryle, Gilbert 200

Saberhagen, Fred 188
Sargent, Pamela 214-16
Schmitz, James H. 42-45, 69, 70, 186
Schwartz, Stephan 170, 172, 200
science fiction 2, and *passim*
Science Fiction and the Quest for Transcendence 37, 190
Science Fiction Hall of Fame 42
The Science Fiction of Mark Clifton 60-61
Science Fiction Research Association 191
"The Science of Psionics" 15
"scientifiction" 21
Scortia, Thomas N. 77

Second-Stage Lensman 28
Sellings, Arthur 112–14, 187
"Sense from Thought Divide" 64
The Several Minds 121–23, 187
Shaw, Bob 210
Sheckley, Robert 215
Sheehan, Daniel P. 201
Sheewash Drive 43–44
Sheldrake, Rupert 215
Shelley, Mary 21
Shepard, Lucius 164–66, 187
The Shining 184
Shoup, Richard 201
Sidgwick, Henry 195
Silverberg, Robert 108, 149–51, 155–58, 164, 180, 185, 187, 214
Simak, Clifford 148, 178, 188, 210–11
Simulacron-3 213
singularity, technological 192
Sinister Barrier 138
sixth sense 3
skeptics 2, 4, 193, 196–98, 204
Slan 9, 18, 32–36, 89, 186
Smith, Cordwainer 212
Smith, E. E. (Doc) 16, 28–31, 80, 178, 186–87
Smith, George 81
Smith, Joseph 50
Smith, Paul 114, 173
"The Snowball Effect" 152
Society for Psychical Research 109, 195
"Something Wild Is Loose" 180, 187
somming 66
Soviet Union 5, 174
space flight 7, 9, 133, 184
The Space Suit Film 184
The Space Vampires 211
special relativity 73
Spielberg, Steven 134
Spinrad, Norman 92, 93
Spiritism 196–97
spiritualism 3, 21
SPOILERS 6
Spottiswoode, James P. 202
Stableford, Brian M. 38, 182, 187, 211
Stapledon, Olaf 11, 19, 21, 24–27, 33, 53, 58, 186
"Star, Bright" 60, 191
Star Fire 173–75, 190
Star Gate Archives 173
Star Gate Program 3, 7, 17, 67–68, 73, 136, 147, 165, 173, 175, 180–205
Star Trek 30, 45, 90, 117
Star Wars 27, 44, 90
Starmaker 11, 19
The Stars My Destination 4, 45, 69, 83–85, 111, 121, 175, 186, 188

steampunk 90
Stevenson, Ian 209
The Stochastic Man 149, 155, 187
Stokes, Douglas M. 202
Straczynski, J. Michael 46
Stross, Charles 212
Sturgeon, Theodore 9, 12, 52–59, 65, 89, 144, 148, 161–63, 184, 186, 192, 210
Supermind 78, 109, 111, 187
supernormals 26–27, 39
Survivor 159
Swann, Ingo 173–78, 190, 206
The Sword of Shannara 92, 145

Talents Universe 130, 187
Targ, Russell 3, 114, 173–74, 206
Tart, Charles 200
Teilhard de Chardin, Pierre 58
"Telek" 92–95, 186
telekinesis (psychokinesis; anomalous perturbation) 3, 13, 52, 67, 87, 94, 171, 187
Telepath 112–14, 187
Telepathist/The Whole Man 89, 118, 123, 136, 187
telepathy 1, and *passim*
That Sweet Little Old Lady 109, 187
theosophy 21
They'd Rather Be Right 59, 62–63, 186
The Thirteenth Floor 213
Thouless, Robert 4
Thrilling Wonder Stories 37
The Ties of Earth 45, 69, 72, 77, 82, 186
Tiger! Tiger! (*The Stars My Destination*) 4, 84
Time and Again 210
Time for the Stars 7, 73, 186
Time Is the Simplest Thing 178, 188
Time Killer 215
The Time Masters 67
Time Pressure 211
time vision 140
Tipler, Frank 211
To Marry Medusa 56–58, 71, 89
To Ride Pegasus 130
To the Stars 74
To Your Scattered Bodies Go 211
transcendence 4, 5, 18, 37, 56–59, 117, 190–93, 210
transvection 167
"The Tree of Time"/*Beyond the Barrier* 78
Trouble with Lichen 113
Tucker, Wilson 18, 67–69, 81, 87, 114, 186
Turing, Alan 66
Twenty Cases Suggestive of Reincarnation 209
2001 7, 107, 189, 209–10, 214
Typhoon (Russian submarine) 203, 204

Ubik 133–38, 187, 213, 214
UFOs 12, 15
United States Army 5
Unknown 138
Utts, Jessica 198, 200

Vance, Jack 55, 92–98, 186
van Vogt, A. E. 8–9, 18, 27, 32, 33–37, 72, 78, 107, 118, 186, 190
Varley, John 212
vatch 45
Vaughn, Carrie 167, 187
Vinge, Joan D. 162, 187
von Lucadou, Walter 201

Wagenmakers, Eric Jan 201–2
Wagner, T. M. 161
Waiting 78, 186
Walker, Evan Harris 215
Walwyn, Isaac 34
Warcollier, René 206
watham 211
Watson, Ian 22, 131, 214, 216
"We *Must* Study Psi" 15
Wells, H. G. 8, 127
Westfahl, Gary 184
What Dreams May Come 216
"What Have I Done?" 60
"What Thin Partitions" 64, 186

"What Will Happen to You When the Soviets Take Over?" 174
When They Come From Space 63
"When You Care, When You Love" 56
White, James 180
White Noise 216
The Whole Man 113, 118
Wiesner, Bertold 4
Wild Seed 159
Wild Talent 67, 87, 114, 186
Willis, Connie 169–73, 187–88, 213
Wilson, Colin 26, 129, 138–41, 176, 186–87, 211
Wiseman, Richard 197–98
"witch burner" gene 132
witches 4, 43, 44, 83, 126
The Witches of Karres 42, 69, 76, 186
"A Womanly Talent" 130
Woodcott, Keith (John Brunner) 115, 187
The World Beyond the Hill 37, 190
The World of Null-A 36
"A World of Talent" 134, 138
Worlds Beyond 99, 100
Wright, Lan 87, 186
Wyndham, John 10, 72–73, 113, 118, 186

Zelazny, Roger 118, 184, 188
Zener cards 146, 171

www.ingramcontent.com/pod-product-compliance
Lightning Source LLC
Chambersburg PA
CBHW021352300426
44114CB00012B/1195